DATE			

An OPUS book

THE PRIMEVAL UNIVERSE

OPUS General Editors

Keith Thomas
Alan Ryan
Walter Bodmer

OPUS books provide concise, original, and authoritative introduction to a wide range of subjects in the humanities and sciences. They are written by experts for the general reader as well as for students.

The
Primeval Universe

JAYANT V. NARLIKAR

Oxford   New York
OXFORD UNIVERSITY PRESS
1988

Oxford University Press, Walton Street, Oxford OX2 6DP

Oxford New York Toronto
Delhi Bombay Calcutta Madras Karachi
Petaling Jaya Singapore Hong Kong Tokyo
Nairobi Dar es Salaam Cape Town
Melbourne Auckland
and associated companies in
Beirut Berlin Ibadan Nicosia

Oxford is a trade mark of Oxford University Press

British Library Cataloguing in Publication Data
Narlikar, Jayant V.
The primeval universe.—(An OPUS book).
1. Big bang theory I. Title II. Series
523.1′8 QB991.B54
ISBN 0–19–219229–9
ISBN 0–19–289214–2 Pbk

Library of Congress Cataloging in Publication Data
Narlikar, Jayant Vishnu, 1938–
The primeval universe. (OPUS)
Bibliography: p. Includes index.
1. Cosmology. 2. Big bang theory. I. Title. II. Series.
QB981.N33 1988 523.1 87–22013
ISBN 0–19–219229–9
ISBN 0–19–289214–2 (pbk.)

Set by Graphicraft Typesetters Ltd., Hong Kong

Printed in Great Britain by Biddles Ltd.
Guildford and King's Lynn

To Mangala

I think that taken as a story of human achievement and human blindness, the discoveries in the sciences are among the great epics and they should be available in our tradition.

Robert J. Oppenheimer

Preface

One of the most exciting frontier areas in science today is the early history of the universe. The discovery of the microwave background radiation in 1965 has swung the opinion of astronomers in favour of the so-called big bang model of the universe. The question that most cosmologists would now like to answer is: What was the universe like soon after its creation in the primeval explosion?

What is 'soon after'? In the late 1940s and the early 1950s George Gamow and his colleagues Ralph Alpher and Robert Herman discussed what the universe must have been like from about 1 second to 3 minutes after the big bang. They found that the universe was very hot, hot enough to synthesize light nuclei in thermonuclear fusion. Out of their work emerged the prediction that today we should see a diffuse radiation background as a relic of that early hot era.

The success of that work has made cosmologists bolder in their speculations. If at the age of 1 second the universe had a temperature of about 10 billion degrees, it must have been even hotter at earlier epochs. What was the nature of matter at those epochs? Here one needs input from fundamental particle physics. For high temperature implies high energy, and one must know how the basic constituents of matter behave at very high energy.

In the 1980s particle physicists have also been more adventurous in their speculations about the fundamental forces that hold matter together. The success of the electroweak theory, which unifies the electromagnetic theory with the weak interaction and whose predictions have been confirmed by accelerator experiments at CERN and the Fermilab, has inspired them to seek even more comprehensive theories, which would unify all the basic interactions of physics. The unification is expected to play a crucial role for matter interacting at very high energies—energies several orders of magnitude higher than can be generated in man-made accelerators, unfortunately. How, then, can one expect to study the actual consequences of unified theories in operation? This is where the current flurry of activity on the interface between particle physics and cosmology comes in. For, at a time sufficiently close to the big bang epoch—around 10^{-37} seconds after it—the constituents of the universe would have been energetic enough to be subject to the full effects of a unified theory. Studies of the earliest epochs should therefore not only tell cosmologists the answer to their question, but also help particle physicists understand the nature of unified theories.

This book describes how cosmologists and particle physicists go about this task, how their joint speculations aim at getting tangible data, and how the hope is to find concrete relics today which will tell us about what went on in the very early universe. Chapters 1–3 describe the cosmological background needed to understand this work, while Chapter 4 is devoted to ideas in particle physics. Chapters 5 and 6 discuss the main scenarios for the early universe. The final chapter is one advising caution, for it is all too easy to get carried away by exciting speculations.

I am grateful to Dr T. Padmanabhan for helpful comments on the manuscript while in preparation. Mr S. T. Satam helped with the illustrations, and Mr P. Joseph with the typing of the manuscript. To both of them my thanks are due for speedy and efficient work. Finally, I must record my deep appreciation to Jean van Altena for polishing the text.

Tata Institute of Fundamental Research   JAYANT V. NARLIKAR
Bombay 400 005, India

April 1987

Contents

1

Introducing the Universe

One day of Brahma elapses when the four yugas *Krita*,
Tretā, *Dwāpar*, and *Kali*, are repeated a thousand
times. One cycle of four yugas takes up 12,000 divine
years. One divine year is equal to 360 million years.

Vishnu Purana, Book 1, ch. 3, vv. 12–15

I am always intrigued by the above quotation, one which is
also found in some other ancient Hindu scriptures. For, doing
the necessary multiplication, we find that the day of Brahma,
the Creator of the universe, is 4.32 billion years, a figure that
is of the same order of magnitude as current estimates of the
age of the universe based on the latest astronomical observations. How did the ancients arrive at this figure?

The rest of the description in the *Vishnu Purana* does not
give any clue. Nor does the spatial description of the universe
shown in Fig. 1.1 come anywhere near to the present picture:
it visualizes the earth and the cosmos as resting on a group of
elephants, who are supported by a giant turtle, who in turn
stands on the divine cobra, the Shesha-nāga.

I have given the above example as one among many which
illustrate how from time immemorial man has been trying to
ascertain the nature of the vast universe he sees around him.
In the remote past these attempts were carried out by
philosophers and religious thinkers. *Cosmology*, the subject
dealing with the origin, structure, and evolution of the
universe is thus as old as the oldest human civilizations. As a
branch of science, however, it is much younger. Indeed, it is
not an exaggeration to say that only after the first quarter of
this century did scientists feel confident enough to tackle the
universe as a physical system subject to the laws of science.
This book is mainly devoted to present-day attempts of

scientists to probe the past of the universe, seeking an answer to the profound question of its origin.

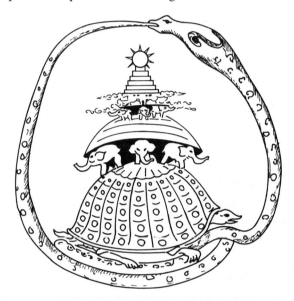

Fig. 1.1. The hierarchical world model envisaged in Hindu mythology. Hindus had other ideas too, including that of *Brahmanda*, the 'cosmic egg' that includes all creation.

From Aristotle to Hubble

Theoretical ideas, however profound, are not by themselves sufficient to raise a subject to the level of a scientific discipline. Theory must make predictions which are testable by experiments or observations. If we were to single out *just one* observation which inspired physicists to take on cosmology as a part of science, it would have to be the discovery of Hubble's law in 1929.

However, as Hubble himself put it in his classic book *The Realm of the Nebulae*, the history of astronomy has been a history of receding horizons. To appreciate the importance of Hubble's discovery, it is necessary first to take stock of the history of astronomy, of the gradual but steady increase in man's understanding of the physical nature of the universe.

While the documented history of human civilizations takes us considerably farther back in time, we will start our ascent of the historical ladder of observational astronomy with Aristotle (384–322 BC). The rungs of this ladder leading to Hubble (1889–1953) are Aryabhata (b. AD 476), Copernicus (1473–1543), Galileo (1564–1642), Kepler (1571–1630), Newton (1642–1727), Kant (1724–1804), Lambert (1728–1777), and Proctor (1837–1888).

Aristotle was a pupil of the great Greek philosopher Plato and the teacher of another famous Greek personality, Alexander the Great. But Aristotle was a distinguished scholar in his own right, and his ideas on the nature of the universe were to dominate not only Greek civilization, but other civilizations in Europe and Asia as well. Indeed, in medieval Europe, Aristotle's ideas were so well established as to become part of religious dogma. The key to Aristotle's ideas lies in his classification of different types of motion.

Aristotle distinguished two types of motion seen in the universe: *natural motion*, which he supposed always to be in circles, and *violent motion*, which was a departure from circular motion, and implied the existence of a disturbing agency. Why circles? Because Aristotle was fascinated by a beautiful property of circles which no other curve seemed to possess. Take any portion of a circle (what we usually call a 'circular arc') and move it anywhere along the circumference: that portion will coincide exactly with the part of the circle underneath it, as shown in Fig. 1.2. (The straight line also has this property, but it can be considered as a circle of infinite radius.)

In the jargon of modern theoretical physics, the above property is one of *symmetry*. A one-dimensional creature moving along the circle will find all locations on it exactly similar, there being no privileged position. In Chapter 2 we will find that present-day cosmologists employ similar symmetry arguments about the large-scale structure of the universe.

If you expose a photographic film for a long time to the light of stars in the night sky, you will develop pictures like Fig. 1.3,

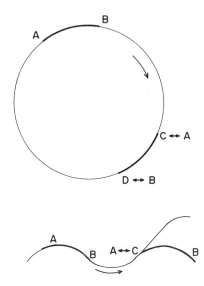

Fig. i.2. Any arc of a circle, like AB, can be displaced and placed congruently anywhere else on the circle, for example, on CD. No other curve drawn in a plane has this property, as can be seen from the curve at the bottom of the figure.

of circular trajectories, as stars rise in the east and set in the west. The sun and the moon also appear to follow circular paths in the sky. We thus see why a prima-facie case existed for Aristotle singling out circular trajectories as nature's favourites.

There was a fly in the ointment, however! A handful of heavenly bodies did not appear to follow circular tracks. Known as 'planets' (meaning 'wanderers' in Greek), these bodies appeared to defy Aristotle's edict of natural motion. Did the planets possess some special power which enabled them to wander at will?

Those who answered this question in the affirmative went the way of the astrologer, who argues that planets not only possess special powers but wield them on human destinies! It is ironical that even though the mystery of planetary motion was fully resolved by Kepler and Newton, and these celestial

Fig. 1.3. This circular trajectories of stars in the sky. Photograph courtesy of the Anglo-Australian Observatory.

bodies were shown to be moving involuntarily under the force of gravitation, belief in astrology is still widespread today.

Greek astronomers who did not go the way of astrology nevertheless missed the chance of possibly discovering the law of gravitation. For, had they considered planetary motion as of the violent kind, they might have been led to wonder about the disturbing agency. Instead, they stuck to natural motion in circles, and to reconcile the manifest lack of circular motion, invented 'epicycles'.

The primary hypothesis behind the epicyclic theory was that the earth is at rest in the universe, and that all heavenly bodies go round it. This theory came to be known as the *geocentric* theory, since everything was viewed with the earth as the centre. The actual motion of bodies round the earth may be circular, as it was found to be for stars, or it may be made up of two or more circles. Thus, in the simplest version, shown in Fig. 1.4, a planet was considered to be moving in a circular

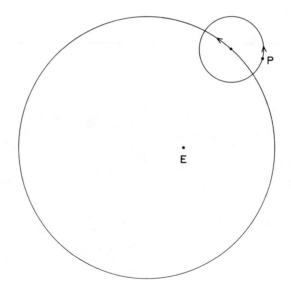

Fig. 1.4. The 'wheels on wheels' type of construction used by the Greeks to describe the motion of a planet, P, around a fixed earth, E.

path whose centre moved on another circular path around the fixed earth. If this description was found inadequate to represent planetary motion and to forecast accurately the position of a planet in the sky, more circles were added to the picture. Ptolemy carried the epicyclic theory to a high degree of sophistication in his classic book *Syntaxis*.[1]

The epicyclic theory was thus no different from the kind of parameter-fitting exercise that goes on in modern times, when resolution of apparent conflicts between observations and a favoured theory is sought by introducing adjustable parameters into the theoretical framework. Such an exercise tells us more about the freshly introduced parameters than it does about the basic hypothesis of the original theory. In fact, as happened with the Greek epicyclic theory, a theory requiring too much patchwork of this sort eventually has to be abandoned.

The first person to take issue with Greek astronomy was the Indian mathematician and astronomer Aryabhata. In his treatise entitled *Aryabhatīya* there is a verse which translates as follows: 'Just as a man moving in a boat downstream sees the trees on the river bank go in the opposite direction, so it is with stars which appear to move in the westerly direction when in fact they are fixed in space.'[2] Aryabhata had thus discovered the fact of the earth's revolution about its polar axis as the cause of the rising and setting of fixed stars. Thus the mystery of Aristotle's natural motion in circles was resolved as far as stellar motions were concerned.

But so deep-rooted were Aristotelian ideas in India in the fifth to eighth centuries that distinguished scholars who venerated Aryabhata in other respects either ignored this argument, or tried to reinterpret it in a way which did not conflict with the 'fixed earth' hypothesis.

A more decisive challenge to the Greek dogma came in the fifteenth and sixteenth centuries, first from Nicolaus Copernicus and then from Galileo Galilei. Copernicus challenged the geocentric theory *in toto*. Not only was the earth rotating about an axis, but it also travelled around the sun, which was fixed in space; or so argued Copernicus in his volume *De*

Revolutionibus Orbium Celestium. In this *heliocentric theory* (in which, as its name implies, everything is viewed with the sun as the centre), epicycles are still needed, but the constructions are simpler than those given by Greek astronomers like Hipparchus and Ptolemy. Moreover, by arguing that the earth and the planets go round the sun, Copernicus paved the way for future theoreticians, who could look to the sun for a possible cause of planetary motion.

Although Copernicus encountered considerable opposition to his revolutionary hypothesis, this was nothing compared to what Galileo had to suffer. Galileo not only supported the heliocentric theory; he also attacked the very basis of the geocentric theory—namely, the ideas of Aristotle. And Galileo's objections to the Aristotelian school were not solely philosophical and conceptual; they were backed by experimental demonstrations. The famous experiment in which he dropped different kinds of bodies from the leaning tower of Pisa is but one among his many demonstrations. Galileo's book called *Dialogue on the Two Great World Systems* about two rival theories is a masterpiece of literary presentation of scientific arguments.

When the Establishment could take no more of this, it reacted in the only way it knew. Galileo was subjected to the Inquisition, for arguing against the established tenets of religion. In the end, he recanted, but privately he retained his convictions about the correctness of the heliocentric theory and his objections to Aristotelian ideas to the end of his life.

Would Copernicus and Galileo have fared any better in modern times, in our so-called enlightened age of science? Today there are no inquisitions, and the Establishment is no longer constituted by religious authorities, but by those responsible for funding scientific projects or refereeing scientific papers in important research journals. One can therefore imagine Copernicus being denied observing time on today's big telescopes, and Galileo having his research grant terminated for propagating unpopular views. Nor would either of them have found it easy to get their research papers published.

It is interesting to note in this context that Tycho Brahe, one of the greatest observational astronomers of his time, shared the geocentric view of the Establishment. He had an excellent observatory at Uraniborg in Denmark, where he conducted observations to establish that the earth is at rest after all! Tycho later moved to Bohemia, where in 1601 he engaged a young assistant by the name of Johannes Kepler to work towards his goal of disproving the Copernican view through detailed observations. Tycho died in 1601, however, leaving behind all the wealth of his observations for Kepler to analyse. Kepler undertook the task in his characteristic meticulous manner, and after some twenty-five years of further observations and data analysis, came to the opposite conclusion, that Copernicus was right after all.

In fact, Kepler took the heliocentric theory considerably further than Copernicus and Galileo. He found, for example, that the planets moved round the sun not in circles (or in circles on circles, and so on), but in ellipses. The elliptical orbits of all the planets had the sun as a common focus. Kepler also discovered empirically the manner in which a planet moved along its orbit. These discoveries were summarized by Kepler in three laws of planetary motion.

Mathematically it is possible to describe the planetary motion in elliptical orbit as made up of a series of epicycles. In principle, an infinite number of epicycles is needed; but in practice, a small finite number gives a good approximation, because the orbits are nearly circular. Had the planetary orbits been considerably more elongated than they actually are, both Ptolemy and Copernicus would have found it hard to sustain the epicyclic construction.

Although Kepler had his own somewhat bizarre interpretation of why planets move according to the three laws (see Fig. 1.5), the correct answer came much later, from Isaac Newton. In 1687 Newton published his masterpiece *Philosophiae Naturalis Principia Mathematica* ('Mathematical Principles of Natural Philosophy'). The *Principia*, as it is known, published in three parts, may be regarded as the first book in theoretical physics as we know it today. Whereas Galileo's famous *Dialogue* had demolished the Aristotelian framework, the

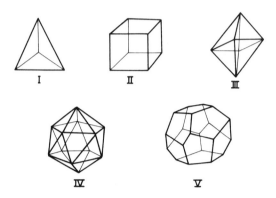

I : Regular tetrahedron

II : Cube

III : Regular octahedron

IV : Regular icosahedron

V : Regular dodecahedron

Fig. 1.5. Kepler believed that the fact that there were *six* planets (the number known in his day) was connected with the geometrical fact that there are only five regular solids (shown above). Nested one inside the other (as shown on page 11), such a frame of solids in the cosmos provided a framework for locating the six planetary orbits. Kepler found this idea compelling for its harmony as created by 'God the Geometor'.

Principia filled the vacuum thereby created by presenting the laws of motion and gravitation. Newton was able to give a mathematical derivation of Kepler's three laws of planetary motion on the basis of his own laws of motion and gravitation.

As we bid goodbye to Aristotle and welcome Newton, let us acknowledge that the Greek philosopher contributed an important notion: namely, that there exist certain basic rules according to which natural phenomena take place. Aristotle's perception of those rules turned out to be incorrect, but the idea that such rules exist was carried over, and has been the guiding light of theoretical physicists to this day.

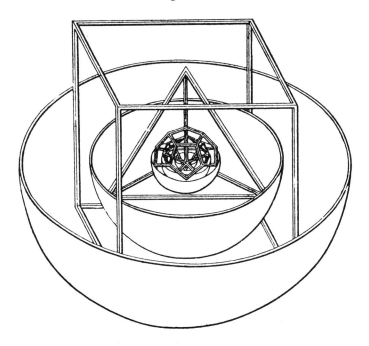

While theory without observational input is sterile, observations without theoretical interpretation do not amount to much. Newtonian theory provided the long-needed interpretation of planetary motion and supplied justification for regarding the heavenly bodies as subject to the laws of science just as much as smaller and nearer systems in the terrestrial laboratory. It was fitting therefore that Edmund Halley, who had been instrumental in getting the *Principia* published, should have applied Newton's laws to comets. Unlike planets, which are almost always seen, comets were noted as occasional visitors to the neighbourhood of the earth and the sun. It was Halley who pointed out that a comet describes a highly eccentric elliptical orbit around the sun. Thus the same comet is seen at regular intervals when it is near the sun, only to disappear as it moves away. Halley regarded sightings of a comet in 1456, 1531, and 1607 as visits of the same comet which, he predicted, would return in 1758. The comet did come as forecast (although Halley was not alive to see it). Known since then as Halley's comet, it has been coming near the sun every 75–6 years, its most recent arrival having been in 1985–6.

The visit of a comet may appear irrelevant to a discussion of the large-scale structure of the universe. But not so! Our theoretical conjectures about the universe have to be rooted in a framework which is well tried and tested. That Newtonian theory can account for cometary motion adds to astronomers' confidence in applying it to even further-away and larger systems. Indeed, it was through such applications that the Newtonian framework contributed significantly to astronomers' understanding of the universe in the eighteenth and nineteenth centuries.

The other major input to astronomy in the seventeenth century was of course the telescope. It was Galileo who first used the telescope for astronomical purposes, and who first appreciated its value in observing remote heavenly bodies. Today we would not be discussing the subject of cosmology had there been no telescopes to give us a view of the universe.

Newton's all-round genius not only encompassed mathematics and theoretical physics; it also went as far as suggesting designs for telescopes. His idea of using a large reflecting concave surface for a telescope, though considered impractical three centuries ago, forms the basis of the large reflectors of today. It is interesting to note that Newton's legacy in mathematics went abroad, to Europe, which produced mathematicians such as the Bernoullis, Euler, Lagrange, and Gauss, while it was his astronomical interests which evoked the more positive response in England.

No one appreciated the usefulness of the telescope more than William Herschel. A busy music master at Bath, Herschel was known for his organ recitals and his huge orchestras. At the age of thirty-five he decided to become an astronomer, largely as a result of night-time reading of books on mathematics and astronomy. Herschel's interest was in observational astronomy, and starting with a small telescope, he eventually

Fig. 1.6. The great reflector of 48-inch diameter by William Herschel played a leading role in stellar astronomy. Photograph courtesy of the Royal Astronomical Society.

went on to build the great reflecting telescope of 48-inch diameter, shown in Fig. 1.6.

Herschel's discoveries are many. On 13 March 1781 he discovered a new planet in our solar system, the one which came to be known as Uranus. This was the discovery which

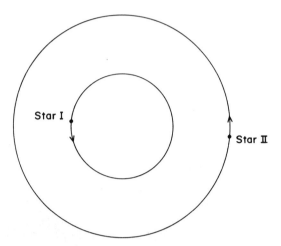

Fig. 1.7. In a binary system two stars go round each other. The observations of binaries provided proof that Newton's law of gravitation applies well beyond our solar system.

brought him immediate recognition, from the Royal Society of London and, more important, from King George III, who became his patron with the grant of a pension for continuing researches in astronomy.

Herschel responded to these honours with prolific work which pushed the observational frontiers of astronomy well beyond the solar system. He noted that several stars in the sky were distributed in pairs, rather than occurring singly. He measured angular separations of some 700 such pairs. For example, the bright star Castor in the constellation Gemini, when seen through a telescope, resolves into two stars of moderate brightness. By comparing his observations with the earlier ones of Bradley, Herschel was able to demonstrate that the stars move round each other much as a planet goes round the sun under Newton's law of gravitation (see Fig. 1.7). This was the first demonstration that Newton's law operates well beyond the solar system.

It was the work of William Herschel and his son John which

led to the first crude picture of our galaxy as a disc-like system of stars encompassed by the white band known as the Milky Way. By examining the distribution of stars away from the sun in all directions, the Herschels concluded that the sun was at the centre of the Galaxy. Thus, although it was known in the nineteenth century that the sun is just a common star which appears to be the brightest object in the sky only because it is the nearest, it still retained the special status of being at the centre of the Galaxy.

This picture of the Galaxy so methodically built up by the Herschels still had two defects, which were not corrected until much later, at the beginning of the present century. But even in the eighteenth and nineteenth centuries there were those who suspected that something was wrong and whose perceptions came remarkably close to the truth as we now know it. The mathematician J. H. Lambert suggested, for example, that the stars in the Milky Way are in motion round a common centre, and that the sun, along with its planets, also moves round this galactic centre.

Lambert also suggested that not all visible objects are confined to our galaxy. In addition to stars and planets, astronomers had also found diffuse nebulae whose nature was not clear. Were they far-away clusters of stars, or were they nearby clouds of luminous gas? William Herschel first opted for the former possibility, but later found that at least some nebulae were like bright clouds around stars, thus suggesting the latter option. Moreover, his counts of the nebulae showed that they tended to stay clear of the Milky Way. He therefore concluded that these nebulae were part of the Galaxy which happened to lie away from the 'disc' spanned by the Milky Way. Lambert, on the other hand, argued that the nebulae were indeed very distant objects, far beyond the Galaxy. But then why did they avoid the disc of the Milky Way?

R. A. Proctor, who supported Lambert's ideas, offered what eventually turned out to be the correct answer: namely, that the apparent avoidance of the galactic disc by these nebulae was due to the obscuring nature of interstellar dust. Because dust absorbs light predominantly in the disc, we are not able to see those nebulae which happen to lie in the direction of the Milky Way.

As late as the decade of 1910–20, however, astronomers still clung to the picture of Herschel. For example, J. C. Kapteyn, using the new technique of photography which had proved such a boon to astronomy, arrived at a model of our galaxy as a flattened spheroidal system about five times larger along the galactic plane than in the direction perpendicular to it. In this model, commonly known as the *Kapteyn universe*, the sun was located slightly out of the galactic plane at a distance of some 2,000 light-years from it (1 light-year is the distance travelled by light in one year, or approximately 10^{18} cm). The sun was thus not too far from the galactic centre, just as Herschel had proposed earlier.

When Kapteyn's work was published in 1920–2, it was already being challenged by Harlow Shapley. In a series of papers published during 1915–19, Shapley studied the distribution of large dense distributions of stars called 'globular clusters'. A globular cluster may contain between 10^5 and 10^6 stars, and can be identified from a distance because of its brightness and distinctive appearance. Shapley found the distribution of globular clusters to be uniform perpendicular to the galactic plane, but not along it. Along the plane they seemed to be concentrated in the direction of the star clouds in the constellation of Sagittarius. Shapley accordingly assumed that the galactic centre lay in that direction, well away from the sun, and estimated that the sun's distance from the centre was 50,000 light-years. The modern estimate of this distance is only about 60 per cent of this value, but the sun does go around the galactic centre, as Lambert correctly guessed.

While Shapley was right in dethroning the sun from its presumed privileged position at the centre of the Galaxy, his distance estimates were too large because he ignored the effects of interstellar absorption, which Proctor had earlier anticipated. Nor did Shapley agree with Proctor's conclusion that most of the diffuse nebulae lay outside the Galaxy. But by the 1920s, the obscuring role of the dust began to be under-

Fig. 1.8. This photograph of the Milky Way in Sagittarius shows clearly the dark regions interspersed with bright ones. The former are due not to any absence of stars, but to obscuration by dust. Photograph courtesy of the Indian Institute of Astrophysics.

stood, and the picture of our galaxy underwent a drastic change. Many stars which were earlier believed to be far away because they looked faint were discovered to be much nearer, their faintness being due to the interstellar dust, as seen in the photograph of Fig. 1.8. Even more important was the conclusion that many of the diffuse nebulae lie far away, well outside the Galaxy. Indeed, it soon became apparent, thanks largely to the work of Hubble, that these nebulae were galaxies in their own right, as large as our own, which are moving away from our galaxy at very large speeds.

Before we proceed to discuss the exciting observations of Hubble, however, we must go back to Immanuel Kant, who two centuries earlier had a cosmological theory which anticipated all this. Kant had argued that nebulae were distant clusters of stars comparable to our galaxy. He had called such nebulae (including our galaxy) 'island universes'—vast populations of stars floating in empty space, far away from one another.[3]

Hubble's Law

By the beginning of the present century, astronomers had two important sources of information regarding typical bright objects in the sky. Photographic plates, more efficiently than the human eye, supplied images of faint objects, while spectrographs analysed the light forming these images into components of different wavelengths. Before proceeding to Hubble's law, let us first note how these two sources of information work.

Even a cursory look at the star-studded night sky tells us that not all stars are equally bright. Indeed, as our eyes grow accustomed to the darkness, we begin to pick out faint stars which had earlier escaped our notice. Likewise, the intensities of images formed on photographic plates also vary from object to object. This variation in intensity is expressed by the astronomer in a quantitative fashion by means of the so-called magnitude scale.

The magnitude scale is logarithmic in character; that is, it measures *ratios* of intensities, rather than *differences*. Suppose we express the intensity of an image as the amount of light energy collected over unit area per unit of time, taking as our unit 10^{-10} watts per square metre per second. Now let the intensities of three astronomical objects A, B, and C be respectively 1, 100, and 10,000 such units. Thus C is as many times (namely, 100) brighter than B as B is brighter than A. The difference in the magnitudes of B and C is thus said to be equal to the difference in the magnitudes of A and B.

By convention, the magnitude scale *ascends* as we go towards fainter objects, with the factor of 100 corresponding to a difference of five magnitudes. So a first-magnitude star is 100 times brighter than a sixth-magnitude star. In terms of the

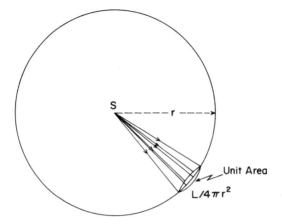

Fig. 1.9. At a distance r from a light source, S, the flux of radiant energy crossing unit area perpendicular to the direction of radiation is $L/4\pi r^2$. All points equidistant from S lie on a sphere whose surface has an area $4\pi r^2$. Thus the total energy leaving the source in unit time is $L = 4\pi r^2 \times (L/4\pi r^2)$. L is the luminosity of the source.

intensity units introduced above, a star of zero magnitude has an intensity of about 252 units.

If all stars were equally powerful radiators of light, the nearer ones would appear brighter than the remote ones; thus, stars of larger magnitude would be further away. On this assumption therefore, magnitudes are indicators of distance. Thus if L is the 'luminosity' of a star at a distance r from us, then the amount of light we receive from the star per unit area per unit time is

$$l = \frac{L}{4\pi r^2}.$$

This is based on Euclid's geometry: the area of the spherical surface of radius r centred on the star is $4\pi r^2$, and, as shown in Fig. 1.9, the star's radiation is supposed to be spread uniformly across it. The star's magnitude, m, is then defined by

$$m = -2.5 \log \frac{l}{l_0},$$

where l_0 is the intensity of a zero-magnitude star. Combining these two equations, we get the relationship between magnitude and distance:

$$m = 5 \log r - 2.5 \log \left(\frac{L}{4\pi l_0} \right).$$

Useful though this formula is, astronomers have learnt to use it with caution, since it has several loopholes. For example, not all stars are equally powerful. An intrinsically very powerful star (with a large value of L) will have a small magnitude even at a comparatively large distance. So, simply from their magnitudes, we cannot decide which of two astronomical objects is the nearer one. Another source of error in the magnitude formula comes from neglecting the absorption of light from the star by intervening dust. We have already seen how the interstellar dust misled nineteenth-century astronomers with regard to the shape and size of our galaxy. If dust is present, it makes the star appear fainter than it would otherwise have looked at the same distance. To correct for this effect, the observed magnitude of the star therefore has to be appropriately *reduced*.

We next turn to the spectroscopic information. As mentioned earlier, the spectrum of an object gives information about how its light is distributed over different wavelengths. We are familiar with the breakup of sunlight when we pass it through a prism. The sunlight is split up into the seven colours of the rainbow, ranging from red at the longest wavelength to violet at the shortest. Although the bulk of the light is distributed over the visible range of 400–800 nm (1 nanometre (nm) $= 10^{-9}$ m) in a continuous manner, the spectrum of an astronomical object may, in addition, contain dark and bright lines which stand out at specific discrete wavelengths.

The dark lines, called 'absorption lines', were first noticed by Fraunhofer in 1814 in the spectrum of the sun (see Fig. 1.10). Known as 'Fraunhofer lines', their true significance became apparent only at the beginning of this century, with the application of quantum theory to the structure of atoms. According to quantum theory, light has the dual nature of a

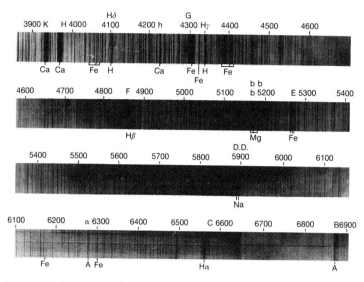

Fig. 1.10. Spectrum of sunlight, showing the dark Fraunhofer lines.

wave and a collection of discrete particles called 'photons'. A light wave of frequency v and wavelength λ travels at speed c. The same wave can be looked upon as made up of tiny packets of energy, called 'quanta', in the form of photons. A photon of frequency v has an energy hv, where h is a universal constant called 'Planck's constant'.

This same feature of discreteness is also present in atomic structure, where it appears in the form of permitted energy levels which an atomic electron can occupy. Moreover, as shown in Fig. 1.11, an electron can move from a level of energy E_1 to another of lower energy, E_2, by radiating a photon of energy $E_1 - E_2$ and frequency $(E_1 - E_2)/h$. The electron can similarly jump up from level E_2 to E_1 by absorbing a photon of this specific frequency. Atoms are in this way 'tuned' to emit or absorb radiation of a specific frequency or wavelength.

An absorption line is produced when incoming hot radiation is absorbed by intervening cooler atoms which are specifically

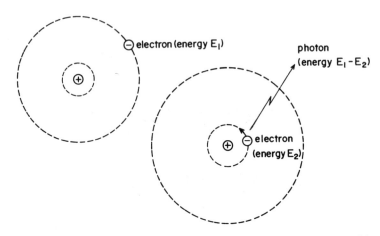

Fig. 1.11. In the upper half, the electron is shown in an outer orbit around the atomic nucleus. In this orbit it has energy E_1. In the lower half of the diagram, it has jumped to an inner orbit, closer to the nucleus, where its energy, E_2, is lower. The energy lost by the electron, $E_1 - E_2$, is carried by the photon which is emitted in the process. Such photons cause a bright emission line of frequency $(E_1 - E_2)h$ to appear in the spectrum of the atoms.

tuned to absorb radiation of that particular wavelength. For example, calcium atoms give rise to absorption lines at wavelengths of 396.8 nm and 393.3 nm, and these are called the H and the K lines respectively. Likewise, bright lines called 'emission lines' arise if hot atoms in a gas radiate at specific wavelengths, again determined by their internal structure.

Emission and absorption lines provide the astronomer with very useful information, since their wavelengths carry the signatures of the emitting and absorbing atoms. Thus the presence of dark lines at wavelengths 396.8 nm and 393.3 nm immediately tells the astronomer that the radiation has passed through a cooler cloud containing calcium.

The spectra of nebulae were first investigated in 1864 by William Huggins by means of visual inspection. Only the continuous distribution was clearly visible; the lines, if any were present, were too faint to be detected. The brightest of the nebulae, M 31, was subsequently investigated in detail. (M 31 is the catalogue number of the galaxy in the compilation made by Charles Messier.) In 1899, J. Scheiner established the presence of absorption lines in the spectrum of M 31, although he was unable to measure their precise wavelengths.

The breakthrough was made in 1912 by V. M. Slipher at Lowell Observatory, using a fast camera attached to the observatory's 24-inch refracting telescope. When Slipher measured the wavelengths of the absorption lines in M 31, he found a somewhat unexpected result. The spectral lines appeared to be displaced from their natural locations, their observed wavelengths being systematically *shorter* than the expected (signature) values. The shift was thus towards the violet end of the spectrum.

By 1925, Slipher's list of such shifts of absorption lines in the spectra of nebulae like M 31 had grown to forty-one, and it had become clear that his first observation of a violet shift for M 31 was more the exception than the rule. The rule indicated a shift in the opposite direction, towards the *red* end of the spectrum. The observed wavelengths of absorption lines in nebular spectra almost invariably *exceeded* the signature wavelengths. Such a 'red-shift' may be defined quantitatively by the following simple formula:

$$z = \frac{\lambda - \lambda_0}{\lambda_0},$$

where λ_0 is the signature wavelength, and λ the observed wavelength.

Physicists already had an explanation for such spectral shifts. First discovered by the Dutch physicist C. Doppler (1803–54) in the case of sound waves, the explanation is equally applicable to light waves. The shifts like those in Fig. 1.12 are expected to occur whenever there is relative motion between the source and the observer. If the relative motion is one of recession, a red-shift is observed; if it is one of approach, there will be a violet-shift. For small shifts, z is simply the ratio of the speed of recession to the speed of light, c.

Edwin Hubble and his younger colleague Milton Humason

Spectral lines blue-shifting

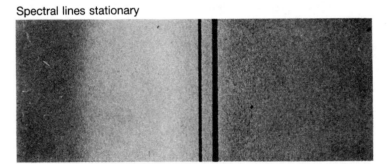

Spectral lines stationary

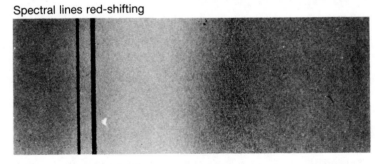

Spectral lines red-shifting

Fig. 1.12. The line in the spectrum shown in the middle is seen shifted towards the 'red' end below and the 'violet' or 'blue' end above. The red-shift occurs when the light source is moving *away* from the observer who took the spectrum. Conversely, the violet-shift arises when the source is moving towards the observer.

carefully analysed the data on spectral shifts. In particular, before deciding whether a distant nebula was approaching or receding from our galaxy, they corrected for our own motion relative to the galactic centre (recall that the earth moves round the sun, which in turn moves round the galactic centre). By 1929 astronomers had made enough progress to be able to make estimates of the distances of many of these faint nebulae. Looking at these distance estimates, Hubble noticed a definite pattern: the larger the distance, the larger the redshift, and hence the speed of recession of the nebula (see Figs. 1.13, 1.14). Hubble expressed this pattern as a velocity–distance relation:

$$V = H \times r,$$

where V is the velocity of a nebula and r its distance. The constant of proportionality, H, in the above is called 'Hubble's constant', and the relation 'Hubble's law'.

It is clear from the above relation that $1/H$ has the dimensions of time. Hubble's own estimate, of about 1.8×10^9 years, turned out to be a gross underestimate, largely because the nebular distance estimates of the 1920s contained several large systematic errors. The present estimates of $1/H$ are in the range $10^{10} - 2 \times 10^{10}$ years. The uncertainty of a factor two reflects the continuing unreliability of astronomical distance measurements, even today.

Another way of expressing Hubble's constant is simply as the ratio of velocity to distance. The astronomer prefers to express velocities in units of kilometres per second and extragalactic distances in units of megaparsecs (1 megaparsec (Mpc) = 1 million parsecs, and 1 parsec (pc) = 3.26 light-years). The Hubble constant is presently believed to lie in the range of $50-100$ km s^{-1} Mpc^{-1}.

The Large-Scale Structure of the Universe

Hubble's 1929 paper in the *Proceedings of the National Academy of Sciences* of the United States marks the beginning of modern cosmology. We will skip over the various observa-

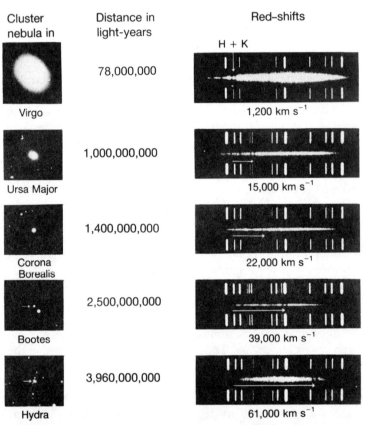

Cluster nebula in	Distance in light-years	Red-shifts

Fig. 1.13. Spectra and photographs of galaxies, showing the Hubble effect: the fainter the galaxy the larger the red-shift of H and K lines of atomic calcium in its spectrum. Palomar Observatory Photograph.

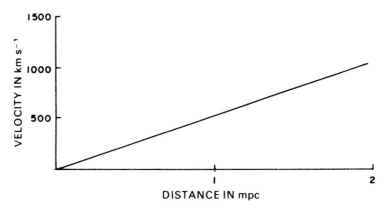

Fig. 1.14. Graph based on Hubble's original observations. The distances estimated by Hubble have since been revised upwards by nearly 5–10 times; but the systematic increase in velocity with distance remains.

tional landmarks since that date, and present the large-scale view of the universe as it is known today.

The velocity–distance relation which Hubble obtained originally from observations of eighteen nebulae has since been determined for a much larger number of galaxies out to distances more than two hundred times greater than those of the nebulae in his original sample. The phenomenon of the red-shift and its relationship to distance appears to hold. So we have prima-facie evidence that other galaxies appear to be rushing away from our own. Has man finally arrived, then, at a situation which singles out his location in the universe as a somewhat special one?

No. A little reflection on the velocity–distance relation will assure us that there is nothing special about our location. Another observer viewing the galactic population from another galaxy will find the same law as Hubble, with all galaxies rushing away from his own.

To see the underlying truth of these observations, imagine the universe as a lattice structure, with galaxies occupying

Table 1.1 Sizes and Masses of Astronomical Systems

Structure	Linear size	Mass
Earth	6.4×10^8 cm (radius)	6×10^{27} g
Sun	7×10^{10} cm (radius)	2×10^{33} g $\equiv M_\odot$
Globular clusters	5–30 pc	$\sim 10^5$–10^6 $M_\odot$
The Galaxy	15 kpc (disc radius)[a]	2×10^{11} $M_\odot$
Clusters of galaxies	1–5 Mpc	10^{13}–10^{15} $M_\odot$
Superclusters	50–100 Mpc	$\sim 10^{16}$ $M_\odot$
The universe	3–6 Gpc (Hubble radius)	$\gtrsim 10^{21}$ $M_\odot$

[a] 1 kpc (kiloparsec) = 1,000 pc.

the lattice points. Suppose now that the entire lattice expands. All points of the lattice will therefore move away from one another, with no particular point occupying a preferred position. This analogy helps us to visualize the oft-repeated statement that the universe is expanding.

The expanding universe exhibits two important properties: there is no privileged position, as we just mentioned, and no preferred direction. An observer brought blindfolded to any location cannot, on removing his blindfolds, tell where he is or in what direction he is looking! Perfect democracy prevails in this universe.

Such uniformity on a large scale does not preclude the existence of hierarchical structures on a smaller scale. Table 1.1 gives some idea of the sizes and masses of such structures, ranging from stars to superclusters. It tells us that the largest distance scale associated with the universe is c/H, which is around 10–20 billion light-years, or 3–6 gigaparsecs (1 gigaparsec (Gpc) = 1,000 Mpc). This is often called the 'Hubble radius'.

Our galaxy is a member of a Local Group of some twenty galaxies. The Local Group itself belongs to a supercluster known as the 'Local Supercluster' shown in Fig. 1.15. It is not known whether even larger systems like super-superclusters also exist.

The task facing the cosmologist today is to interpret this vast universe as a physical system subject to the known basic laws

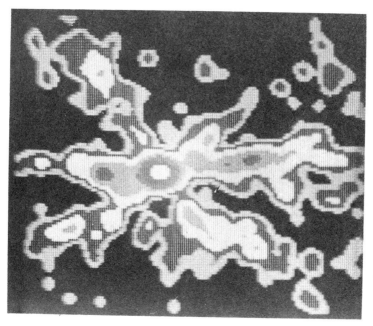

Fig. 1.15. A computer picture of the Local Supercluster, which is typical of the largest discrete structures known to exist in the universe today.

which appear to govern the behaviour of matter. Given the information that this system is expanding, we would like to know what it will do in future and, more important, what it was like in the past. For our speculations regarding the future of the universe will be reliable only if our reading of its past history is based on solid foundations. The rest of the book is therefore devoted to a survey of scientific attempts to probe the past of the universe.

2

The Big Bang

Philosophy is written in that great book which ever lies
before our gaze—I mean the universe—but we cannot
understand if we do not first learn the language and
grasp the symbols in which it is written.

Galileo Galilei

How does one react when confronted with the awesome
evidence that the universe is expanding? Reactions have been
many and varied. There are those who are totally at a loss to
comprehend the magnitude of the phenomenon, persons to
whom distances of hundreds of millions of light-years and
speeds approaching the speed of light do not begin to convey the grandeur of what is being observed. At the opposite
extreme are those who do appreciate the largeness of the
universe, but who confess that astronomy has finally led us to
an event (the big bang) which does not fall within the scope of
normal scientific enquiry.

Between these extremes are reactions which are expressed
through questions like these: What is the universe expanding
into? Is there an infinite void beyond the observed multitude
of receding galaxies? When did the expansion begin? When
will it end, if ever? Will it be followed by a contraction?

These questions and others like them fall within the scope
of rational scientific enquiry. Indeed, as we saw in Chapter 1,
Hubble's discovery of 1929 acted as the stimulus for the
consideration of cosmology as a branch of science. Scientific
cosmology has since come a long way, thanks both to the
increasing sophistication of astronomers' techniques for observing the universe and to the growing confidence of theoreticians in the laws of physics which govern the behaviour of the
universe. The twentieth century has thus witnessed the trans-

formation of cosmology from a speculative philosophical enterprise to a scientific discipline.

Which Laws of Physics are Relevant to Cosmology?

The tenets which have guided the progress of science since the days of Galileo and Newton may be summarized as follows:

(1) All observed natural phenomena are subject to definite scientific laws.

(2) Diverse and apparently unrelated phenomena may be explicable by the same basic law.

(3) The number of basic laws in science is very small, and even these laws may eventually turn out to be many facets of a *single* basic law of nature.

Tenet 1 brings all natural phenomena within the scope of science; tenet 2 stresses the universality of scientific laws; while tenet 3 aims at economy and unification in the scientific description of nature. Twentieth-century science tells us that the behaviour of 'non-living' matter is, or should be, explicable in terms of four fundamental interactions in physics.

It is arguable whether these basic laws of physics will also ultimately account for the phenomenon of 'life'. The gap between molecular biology and physics is beginning to close, but it is not yet clear whether new basic inputs in physics are needed to explain life.

All astronomical observations of a cosmological nature are of non-living matter, and the hope is that they can be interpreted within the present set-up provided by physics. However, this is not to deny that biological considerations are relevant to cosmology; indeed, towards the end of this book we will indulge in some cosmo-biological speculations. The bulk of our description, however, will be within the framework of physics as we know it today.

Let us briefly examine the basic physical laws known today, and see which of them might be relevant to our understanding of the phenomenon of the expanding universe. As indicated above, there are four basic interactions of physics: the electromagnetic interaction, the weak interaction, the strong interaction, and gravitation. I have listed them not in the order in which they were discovered, nor in the order of their intrinsic strength as physical forces. Rather, the order reflects the *degree* of our understanding of the character of the interaction.

The electromagnetic force has been the one most thoroughly investigated, both theoretically and experimentally. It was discovered as two sets of isolated phenomena, one involving the force of repulsion (attraction) between like (unlike) electric charges, the other involving the force between magnets. Experimental studies by Coulomb, Volta, Ampère, Faraday, and Maxwell in the seventeenth and eighteenth centuries led to the unification of the electric and magnetic forces. The advent of quantum theory in the twentieth century further revealed the richness of the electromagnetic interaction on the small scale of the atom.

Although important on the atomic scale of 10^{-8} cm, the electromagnetic force is long-range. This property is typically expressed by formulae like the following:

$$F = \pm \frac{q_1 q_2}{r^2}$$

and

$$P = \frac{1}{4\pi} \frac{e^2 a^2}{c^3 r^2} \sin^2 \theta.$$

The first expresses Coulomb's law, that the force, F, of repulsion ($+$) or attraction ($-$) between like or unlike charges q_1 and q_2 varies inversely as the square of their distance apart, r. The second gives the power radiated by a charge e radially away from it as measured at a point located at distance r from it in a direction making an angle θ with the direction of the acceleration, a, of the charge. Historically speaking, the first formula marks the beginning of our understanding of electromagnetism, while the second expresses our recognition of the unity of electricity and magnetism in the phenomenon of electromagnetic radiation.

The latter also tells us why the entire subject of cosmology owes its origin to electromagnetic theory. All the observations reported in Chapter I were of electromagnetic radiation (whose visible manifestation is light). But for this phenomenon we would not know about stars, galaxies, clusters, and superclusters. We also recall that the spectral resolution of radiation into waves of different wavelengths and the quantum electromagnetic effects of spectroscopy formed the basis of Hubble's law, which is what led to the concept of the expanding universe.

But having said this, we still have to dismiss electromagnetic theory from our considerations of the *dynamics* of the expanding universe. For the electromagnetic *forces* of which the first formula gives an example require large-scale conglomerations of electric charges and/or large-scale electric currents. There is no evidence of these phenomena to date; the large masses of relevance to cosmology—the galaxies—appear to be electrically neutral. So we must look elsewhere for the dominating fundamental force of cosmology.

Both the weak and the strong interactions were discovered in this century. The weak force is at the root of the phenomenon of *beta decay*, wherein a free neutron decays into a proton, an electron, and an antineutrino:

$$n \rightarrow p + e^- + \bar{\nu}.$$

The neutron is electrically neutral, the proton is positively charged, and the electron is negatively charged. Thus the electric charge would be conserved even if there were only two products (p, e$^-$) of the beta-decay reaction. It is necessary, however, to introduce a third electrically neutral product, the antineutrino, to conserve the *spin* and the *lepton number*. Spin is an intrinsic property of all the above particles, and it gives each of them an angular momentum of $\frac{1}{2}\hbar$. The lepton number keeps count of how many *light* particles, electrons and neutrinos, take part on each side of the reaction, with particles counted positively and antiparticles negatively.

Leptons partake in both the electromagnetic and the weak interactions. Although the latter does not require that the interacting particles be charged, an exchange of charge often takes place between weakly interacting particles. These circumstances led to the suggestion that the two interactions might in fact be related, being aspects of a single *electroweak* interaction, just as electricity and magnetism had turned out to be aspects of a single electromagnetic interaction. A theory unifying the electromagnetic and weak interactions was successfully formulated in the 1960s by Abdus Salam, Steven Weinberg, and Sheldon Glashow. We will have more to say about the electroweak theory in Chapter 4.

While a free neutron decays in a matter of $\sim 10^{13}$ seconds in the above fashion, it remains stable in the company of protons within the nucleus of an atom. A typical atomic nucleus, X, containing N protons and $M - N$ neutrons has atomic number N and atomic mass M. It is symbolically written as $^M_N X$. For example, the helium nucleus $^4_2 He$ has two protons and two neutrons. Protons and neutrons are together called 'nucleons'. They belong to a family of relatively heavy particles called 'baryons'.

Why does a neutron remain stable within a nucleus? How is a nucleus able to hold together so many positively charged particles despite their mutual electrical repulsion? The answers to these questions lie in the existence of the *strong* interaction, which, at the nuclear level, exceeds in strength both the weak and the electromagnetic interactions. It is this reaction which provides the binding force that keeps the nucleus as a whole stable. It acts selectively, on baryons only.

As we shall see in later chapters, both the strong and the weak interactions contain clues to the very basic cosmological questions regarding the origin and form of the matter of the universe. For example, what is the composition of the universe in terms of different atomic nuclei like hydrogen, deuterium (heavy hydrogen), helium, carbon, and so on? More fundamentally, is the universe predominantly made of matter, or does it contain an equal amount of antimatter? The cosmologist cannot ignore the weak and the strong interactions in seeking answers to these questions. As we shall discover later, the key to many of these questions may lie in current attempts to unify all the basic interactions under one banner. For example, so-called grand unified theories (GUTs) seek to unite the strong and the electroweak interactions.

But none of these interactions serve our present goal of trying to understand the phenomenon of the expanding universe. For the weak and the strong interactions are of extremely short range: their effects are felt only at distances of the order of $<10^{-12}$ cm. This is a far cry from the cosmological distance scale of 10^{28} cm! Again we must look elsewhere for the solution to our problem.

And that leaves us with gravitation, the force whose existence was the first to be recognized, but whose real nature still eludes us at the fundamental level. (Even Newton refused to be drawn into speculations about why the force of gravity exists and why it follows the inverse square law. 'Hypotheses non fingo' ('I do not frame hypotheses') was his comment!) Fortunately for the cosmologist, gravitation marks the end of his quest. For this force, though imperfectly understood, contains all the basic ingredients to make it highly relevant to the cosmological problem of the expanding universe. Let us look at its qualifications briefly, using its formulation in Newton's inverse square law.

Enunciated in 1687 in *Principia*, Newton's law of gravitation may be expressed in the following formula:

$$F = -G\frac{m_1 m_2}{r^2},$$

where F is the force of attraction between two particles of masses m_1 and m_2 separated by a distance r, and G is a constant of nature, whose value is about 6.668×10^{-8} cm^3 s^{-2} g^{-1}.

We will now investigate why gravitation is something like an ugly duckling for the laboratory physicist, concerned as he is with atoms, molecules, or their interiors. Notice first that both Newton's law and Coulomb's law have the same inverse square of distance form. But the atomic physicist ignores the former and utilizes the latter. Why?

The answer is simple. Let us compare these forces for the

simplest atom, that of hydrogen, which contains only two particles, one proton and one electron. We find that the Coulomb force of attraction between them is

$$F_C = -\frac{e^2}{r^2},$$

where r is their separation and $+e$ and $-e$ their respective charges. The Newtonian gravitational force of attraction between them, on the other hand, is

$$F_N = -G\frac{m_p m_e}{r^2},$$

where m_p and m_e are the masses of the proton and the electron respectively.

With $e = 4.8 \times 10^{-10}$ e.s.u., $m_p = 1.67 \times 10^{-24}$ g, and $m_e = 9.11 \times 10^{-28}$ g (and the value of G given above), it is not difficult to verify that the ratio of F_C to F_N is very large, on the order of 10^{40}. Thus, there is no practical advantage to including F_N in the discussions of atomic physics.

The ugly duckling of gravity acquires a swan-like majesty, however, in the cosmic setting. As we mentioned earlier, the Coulomb electric force ceases to be of any import among large astronomical objects like stars and galaxies. The Newtonian gravitational force builds up to enormous strength, on the other hand, simply because it is additive in terms of the accumulation of matter, and an astronomical object has a large number of particles. A typical star like the sun has about 10^{57} particles like the proton, while the number in a typical galaxy is as high as 10^{68}. Moreover, gravity is also a long-range force, and its effect, although diluted by the inverse square law of distance, extends over hundreds of thousands of light-years.

As a measure of how strong the gravitational force of attraction is at the intergalactic level, let us compute the *binding energy* of two galaxies, each of mass 10^{11} M$_\odot$, separated by a distance of 50 kpc. (M$_\odot \cong 2 \times 10^{30}$ kg.) This binding energy, B, tells us the amount of work that must be put in to tear these galaxies apart from each other's gravitational attraction. The answer is

$$B \cong G \times \frac{(10^{11} \, M_\odot)^2}{50 \, \text{kpc}} \cong 2 \times 10^{59} \, \text{ergs.}$$

To grasp the largeness of this number, compare it with the energy released in a megaton hydrogen-bomb, a mere 4×10^{22} ergs, or with the energy radiated by the sun over the last billion years (assuming its present rate), 1.2×10^{50} ergs.

It is evident, therefore, that gravity is the force we have to reckon with on the cosmic scale, and as far as our present knowledge goes, it is the *only* force relevant to our discussion of the dynamics of the expanding universe.

The General Theory of Relativity

Having decided that it is the gravitational interaction which matters on the cosmological scale, we now turn our attention to the gravity theory which is usually considered appropriate to cosmology.

The gravity theory which played a key role in astronomy right from its inception is of course embodied in Newton's inverse square law. This theory successfully explained Kepler's laws of planetary motion; it accurately predicted the periodic arrivals of comets like Halley's; it was responsible for the discovery of the planet Neptune; it is used in the study of the structure and evolution of stars like the sun and in discussions of the dynamics of clusters of stars in the Galaxy. At first sight, therefore, the Newtonian theory of gravity appears to have a strong case for providing the framework of the expanding universe.

However, there are chinks in the armour of Newton's theory which make it suspect from the cosmological point of view. First, we note that the theory assumes that the effect of gravity is instantaneous over any distance, however large. This assumption runs counter to the now well-established principle of Einstein's special theory of relativity that *no* physical effect can be transmitted with a speed exceeding the speed of light. The discrepancy between this principle and the Newtonian law of gravity can be ignored in the astronomical situations described above, because the distances and time spans involved are comparatively small. In cosmology, however, we are dealing with distances of several billion light-years and time spans of several billion years. Here we can no longer sweep the discrepancy under the rug.

The laws of Newtonian dynamics are also suspect when we are discussing objects moving at speeds which approach the speed of light. Since we anticipate that in our description of the expanding universe, such speeds will be encountered, it is therefore necessary to look for a theory which incorporates the essential (and well-tested) features of Newtonian dynamics and gravity at slow speeds and in the astronomical situations described above, but which poses no conflict with Einstein's special theory of relativity at high speeds.

It was Einstein himself who proposed such a theory, a decade after his presentation of special relativity in 1905. Known as the 'general theory of relativity', it provides the ideal framework for a mathematical description of the expanding universe.

In formulating the general theory of relativity in 1915, Einstein was aware of the conflict which exists between the Newtonian law of gravitation and the special theory of relativity. The conflict was, in fact, worse than we have indicated so far; for it also implied that in the presence of gravity the special theory of relativity itself breaks down. Let us try to understand the nature of this objection, for in its resolution lies the basic feature of general relativity.

The special theory of relativity refers to space–time measurements made by different observers, all of whom are moving under no forces. Such observers are called 'inertial' observers, and their motions are considered to be unaccelerated—that is, they all move with uniform velocities. Any one observer can consider himself at rest and the others moving relative to him with uniform velocities. The formal structure of physical laws is assumed to be the same as observed by all inertial observers. It should therefore be impossible to detect the 'absolute motion' of any observer by any experiment.

Since Maxwell's electromagnetic theory should have the

same form for all inertial observers, according to this principle, it follows that the speed of light (or of an electromagnetic wave in general) should be the same as measured by all such observers. But this is clearly impossible if we follow the Newtonian formula for relative velocities, according to which an observer approaching a light source should measure a larger value for the velocity of light than an observer stationary with respect to the light source.

Thus it became necessary to revise the Newtonian concepts of space–time measurements and replace them by new ones. The difference between the old and the new rules connecting the measurements of two inertial observers in relative motion is illustrated by the following simple example. Suppose two such observers, O and O′, use the rectangular Cartesian coordinates x, y, z and $x′, y′, z′$ for their spatial measurements such that their origins, as well as the coordinate axes, coincide at some instant. Let O and O′ start measuring their times t and $t′$ from this instant, and suppose that O finds O′ moving with speed u in the positive x-direction. Then O′ will find O moving with speed u in the negative $x′$-direction. The situation is illustrated in Fig. 2.1.

The old rules connecting the x, y, z, t coordinates of any event K measured by O with the $x′, y′, z′, t′$ coordinates of the same event measured by O′ are given by the *Galilean transformation*:

$$x′ = x - ut, \ y′ = y, \ z′ = z, \ t′ = t.$$

The new rules of special relativity are expressed by the *Lorentz transformation*:

$$x′ = \frac{x - ut}{\sqrt{\left(1 - \dfrac{u^2}{c^2}\right)}}, \ y′ = y, \ z′ = z, \ t′ = \frac{t - \dfrac{ux}{c^2}}{\sqrt{\left(1 - \dfrac{u^2}{c^2}\right)}},$$

where c is the speed of light.

We will not go into the details of special relativity here. Suffice it to say that measurable differences between the two

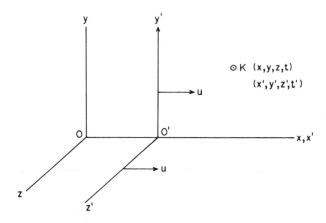

Fig. 2.1. The reference frames of two inertial observers, O and O', in uniform relative motion in the common Ox, Ox' direction. The rules connecting (x, y, z, t) coordinates of a typical event K in the frame of O to (x', y', z', t') coordinates in the frame of O' differ in special relativity from the rule used in Newtonian mechanics.

transformation laws arise as u approaches c, and that observations of fast-moving particles have confirmed the validity of the Lorentz transformation.

But let us return to the conflict which develops between special relativity and gravitation. The conflict can be expressed by this simple question: Where are these inertial observers in the real universe if it contains the all-pervading and ever-present force of gravity? The answer is nowhere! For all observers are subject to the force of gravity, and therein lies the conflict. Special relativity appears to be formulated in terms of observers who do not exist, since all observers are subject to the force of gravity.

The statement in the preceding paragraph about the confirmation of special relativistic formulae must therefore be only approximately true, and only then for situations in which the force of gravity is small enough to be negligible. While such approximations are permitted in the experimental verification of a theory, they have no place in its theoretical formulation, which must be *exact*.

Indeed, the non-existence of truly inertial observers in real life is linked with the all-pervading nature of gravity. While we can 'remove' the electric field from a given region of space, we cannot do the same for gravity. Taking his cue from this unusual property of gravity, Einstein argued that we should look upon gravity not just as an ordinary force, but as a permanent and ambient feature of space and time. And he went on to give geometrical expression to this concept.

Geometry deals with rules of measurement of distances and angles, rules which were first formulated by Euclid (*c*.300 BC). It is Euclid's geometry which we study at school and learn to apply in practice. The Lorentz transformation tells us how to link space measurements with time measurements, and we can extend the rules of Euclid's geometry for spatial measurements to include temporal measurements also. It was Hermann Minkowski who, in 1905, showed how this can be done, and how we can talk of a 'geometry of space–time' in the same way as we talk of geometry of space alone. The Lorentz transformations thus relate the geometrical measurements made by different inertial observers.

But, as we saw just now, Minkowski's geometry is unrealizable in practice, because gravity forbids the existence of inertial observers who would use such a geometry. Einstein therefore had the idea of looking for a *new* geometry which would automatically allow for *real* observers subject to the force of gravity.

To the uninitiated, it may come as a shock to discover that geometry is not unique, that other geometries, which are different from Euclid's, do exist as perfectly respectable, logically self-consistent subjects of study. It took mathematicians nearly two millennia after Euclid to appreciate this fact, and non-Euclidean geometries emerged only during the eighteenth century. Even then, these geometries were considered more as abstract intellectual exercises than as being of any relevance to the real world.

It was to these geometries that Einstein turned in order to express his new idea in a quantitative manner. But before describing Einstein's use of non-Euclidean geometries, let us see how these geometries provide alternative theorems to those of Euclid's geometry.

Suppose that the earth is perfectly spherical, and that it is inhabited by 'flatmen', creatures who are two-dimensional, with no sense of 'height'. Crawling on the terrestrial surface, these creatures identify 'straight lines' as lines of shortest distance by stretching threads across the surface between any two given points. To them such a line appears straight as they move along it, in the sense that their arrival and departure directions at any point on the line have zero angle between them.

Within this definition, the flatmen find that all straight lines intersect, and that moving along any straight line they eventually return to their starting-point. They also discover that the three angles of any triangle they draw on the earth do not add up to two right angles as Euclid's theorem decrees. Rather, the sum of the three angles always *exceeds* two right angles. Fig. 2.2 illustrates a situation where the sum is *three* right angles.

Confronted with this discrepancy, an intelligent flatman will soon discover its cause. He will point out that one of Euclid's basic assumptions of geometry is not true as far as his world is concerned. This assumption is known as the 'parallel postulate', and it states that, 'Given a straight line *l* and a point *P* outside it, we can draw *one and only one* straight line through *P* parallel to *l*.' (Reminder: parallel lines do not meet even if they are stretched out indefinitely.) To us Euclid's postulate appears reasonable; but not to the flatman. He will argue that in his world there are no parallel lines, since all straight lines intersect. And if he examines the proof of Euclid's triangle theorem, he will discover that it makes use of the parallel postulate.

A basic assumption like the parallel postulate is in principle unprovable in terms of the other postulates. Not realizing this, many mathematicians attempted to prove the parallel postulate in the Euclidean framework. Not surprisingly, they failed. As we see in the above example, we could have a different geometry, in which the parallel postulate is changed

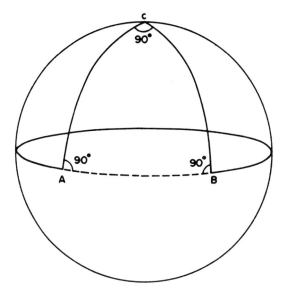

Fig. 2.2. This triangle on the surface of the earth has its three angles adding up to three right angles.

to: 'Given a straight line *l* and a point *P* outside it, we *cannot* draw *any* straight line through *P* parallel to *l*.'

Likewise, we get another perfectly reasonable geometry if we depart from Euclid's parallel postulate in the opposite direction and say: 'Given a straight line *l* and a point *P* outside it, we can draw *two or more* lines through *P* parallel to *l*.'

Let us call Euclid's geometry type I, the geometry on the spherical surface type II, and the geometry of the third kind described above type III. Type III geometry operates, for example, on the surface of a saddle.

There is another practical way in which we can distinguish between these three types. Take a sheet of paper and lay it across a flat surface. The paper will cover the surface smoothly. Try to do the same to a spherical surface and you will find that in order to cover it, you have to allow wrinkles to develop in the paper, indicating that near any given point on the surface the area of the paper is greater than the area that

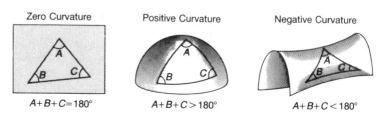

Zero Curvature   Positive Curvature   Negative Curvature

$A+B+C=180°$ $A+B+C>180°$ $A+B+C<180°$

Fig. 2.3. The plane surface is an example of type I space, wherein Euclid's geometry holds; on the sphere, geometry of type II holds; whereas on a saddle-shaped surface, the geometry is of type III.

you are trying to cover. The reverse happens on the saddle surface. The area of the paper turns out to be insufficient to cover the surface near any point on it, and the paper is torn.

Mathematicians express the above result by the notion of *curvature* of a surface. In the above example, the flat surface has zero curvature, the sphere has positive curvature, and the saddle has negative curvature. We also see that geometries of types I, II, and III apply respectively on surfaces of zero, positive, and negative curvature. As shown in Fig. 2.3, the triangle theorem of Euclid holds only in case I. (Caution: Not all surfaces that appear curved to our eye have curvature as per the above definition! Repeat the paper experiment on the curved surface of a cylinder and you will appreciate the truth of this remark.)

We can thus link departure from Euclid's geometry to the existence of curvature, positive or negative. Our examples of two-dimensional surfaces can naturally be extended to higher dimensions and to space–time itself. When we do this, we come close to appreciating the basis of general relativity.

Let us now look at an experiment from the different points of view of Newton and Einstein. Imagine a cannon-ball being fired at an angle of 70° to the vertical direction. How will the ball travel? If we neglect air resistance, Newtonian dynamics tells us that the ball will move in a parabolic curve, rising to a maximum height and then descending to the ground, while travelling all the time with a uniform speed in the horizontal

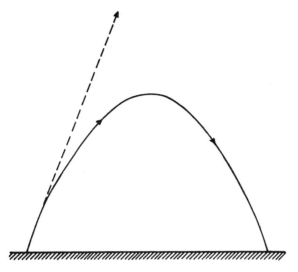

Fig. 2.4. The dashed trajectory shows the path which would be taken by a cannon-ball in the absence of gravity. The continuous parabolic trajectory illustrates the path actually followed by the ball under the influence of the earth's gravity.

direction. How does the parabolic trajectory arise? Had there been no force acting on the cannon-ball, it would have travelled with uniform speed in the direction in which it was fired—that is, in the direction making an angle of 70° with the vertical. It deviates from this straight path only because it is 'pulled down' by the force of the earth's gravity, and Newtonian gravity provides the force formula. (See Fig. 2.4.)

Einstein's approach to this problem is entirely different. First, we note that the statement 'Had there been no force acting on the cannon-ball, it would have travelled with uniform speed in the direction it was fired' is unverifiable, since gravity is always present. In general relativity allowance is made for this situation by modifying the space–time geometry. Thus we say that the cannon-ball moves in a space–time whose geometry is non-Euclidean and non-Minkowskian. Also, since there are no forces acting on the ball (remember, gravity has already been allowed for through its effect on geometry), the first law of motion operates; so the cannon-ball moves with uniform speed in a straight line.

How can non-uniform motion along a manifestly parabolic path be construed as 'uniform motion in a straight line'? The answer to this question, of course, is that because the rules of geometry have changed, so have the notions of straight line and the measurement of speed. According to the new rules, the actual motion of the cannon-ball is uniform and in a straight line.

So, according to general relativity, to handle any problem of gravitational interaction, we must first know how the space–time geometry is modified. Einstein's field equations of general relativity tell us (in principle) how to relate the parameters of space–time geometry to the presence of matter and energy in it. Symbolically we may write:

matter/energy ‹ › gravitational effect ↔ curved space–time.

The arrows go both ways. Just as we can say that matter and energy, through their gravitational effect, modify the space–time geometry from the Minkowski type, so also we can assert that departures from Minkowski space–time geometry imply the existence of matter and energy. The actual field equations are a formidable set of ten second-order partial differential equations (with the space–time coordinates as independent variables), whose solution tells us not only about the geometry of space–time but also about how matter and energy are distributed in it at all times.

The Schwarzschild Solution

Exact solutions of Einstein's field equations have been few and far between. The first problem to be solved following the creation of the general theory of relativity in 1915 was the determination of space–time geometry around a spherical mass situated in otherwise empty space. This problem was solved in 1916 by Karl Schwarzschild.

Although our main interest here is to know how to solve the

problem of the expanding universe containing billions and billions of receding galaxies, we will spend a little time studying Schwarzschild's solution. For this solution played two important roles: first, it provided the link between Newtonian theory and general relativity; and second, it led to new predictions which could be tested by astronomical observations, predictions which inspired general confidence that general relativity was on the right track.

Intuitively we might expect space–time geometry to depart significantly from Minkowskian geometry in the neighbourhood of a spherical mass where its gravitational effect is the strongest. Further, since Newton's inverse square law leads us to expect a diminishing effect of gravity as we move away from the mass, the geometry should approach the Minkowskian form more and more closely at points located farther and farther away from it. In technical jargon, the space–time under such conditions is said to be 'asymptotically flat'.

Schwarzschild's solution confirms this intuitive picture. The words 'far' and 'near' also receive a quantitative meaning in the following way. Suppose that in the Newtonian theory a spherical object O has a mass M. Then at a radial distance r from its centre, its force of attraction on an object P of unit mass would be $F = -GM/r^2$ (the negative sign denoting attraction).

This law can be tested by examining the trajectory of P as it orbits around O. (Historically, planetary motion confirmed the inverse square law.) How would this trajectory compare with the trajectory of P in the Schwarzschild solution? From the example of the cannon-ball, we know how such a trajectory is to be computed—namely, according to the rules of Schwarzschild's geometry. It turns out that this trajectory agrees with the Newtonian trajectory if r is large compared to $R_S = 2GM/c^2$, where R_S is the 'Schwarzschild radius', the only parameter necessary to quantify Schwarzschild's geometry. The larger r is compared to R_S the closer is the agreement between Newton and Einstein. The departure of the actual geometry from the Minkowskian form becomes larger and larger as r approaches R_S. In this 'near zone' the lack of agreement between the theories of Newton and Einstein becomes pronounced.

Let us examine this relationship in the context of our solar system. The mass of the sun is 2×10^{33} g, and the value of R_S for the sun is a mere 3 km. The sun's actual radius is about 700,000 km. Thus any exterior piece of matter (for example, a planet, a comet, or an asteroid) will always be in the 'far zone', and its trajectory computed according to Einstein's theory will be very nearly the same as its trajectory given by Newton's theory; hence the apparent success of Newtonian gravity in explaining Kepler's laws of planetary motion and the motions of the various objects in the solar system.

The phrase 'very nearly' in the paragraph above does not rule out minute, but detectable, differences between the predictions of these two theories. For example, according to Newton's law the orbit of a planet should be an ellipse with the sun as one of its foci. The orbit computed according to Einstein's rule, on the other hand, is an ellipse which is very slowly rotating. Thus the direction from the sun to the perihelion, the point in the orbit nearest to the sun, slowly rotates in the same sense as the motion of the planet. This is known as the 'precession of the perihelion'.

This discrepancy, illustrated in Fig. 2.5, is naturally largest for Mercury, the planet closest to the sun. The calculated rate of precession of Mercury's perihelion is 43 seconds of arc per century. Or, to put it differently, the sun–perihelion direction will make a complete round in about 3 million years. Small though it is, this effect had indeed been observed, and had worried even the astronomers of the last century. Its explanation by general relativity was therefore considered a great success of the theory.

The observation which contributed most, however, to making general relativity and Einstein himself popular the world over was that of the bending of light. The British astronomer A. S. Eddington was mainly instrumental in initiating this observation. Its rationale is simple: light travels in straight lines, and since straight lines in a curved space–time differ from those in Minkowski space–time, light tracks passing

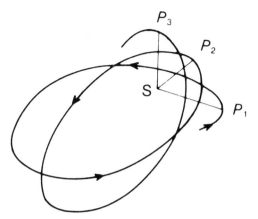

Fig. 2.5. An exaggerated illustration of how the perihelion of Mercury's orbit systematically shifts through positions P_1, P_2, P_3, ... in relation to the sun, S.

near massive objects should be different from those not passing near massive objects.

In practice we should be able to observe this effect as the sun crosses the line of sight to a background star. When the sun is not anywhere near the line of sight, light from the star travels to us as in a Minkowski space–time. Just before crossing, however, the light ray has to pass through the Schwarzschild type of space–time, and so its track should change. The star image should therefore show a change in direction. The maximum predicted effect is a shift of 1.75 seconds of arc in the star's direction. The scenario is illustrated in Fig. 2.6.

Realizing that a total solar eclipse would be required to conduct such an experiment, Eddington proposed making observations during the 1919 eclipse. Observations by two teams, one in Sobral in Brazil, the other on the island of Principe in the Gulf of Guinea, confirmed Einstein's prediction. The date, 6 November 1919, of the joint meeting of the Royal Society and the Royal Astronomical Society in London at which the result was first announced marks the historic epoch when Newtonian gravity was dethroned by Einstein's general theory of relativity.

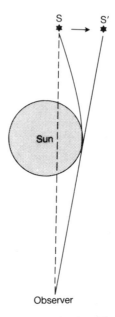

Fig. 2.6. The bending of light by gravity. The actual effect (the shift in the star's direction) has been confirmed.

Relativistic Cosmology

Our search for a physical framework for cosmology is over. Having convinced ourselves that, of all the known fundamental interactions of physics, gravity is the one most relevant to cosmology, and that general relativity provides a more satisfactory theoretical description of gravity than the Newtonian framework, we now proceed to use relativity to construct a mathematical model of the expanding universe.

What is a mathematical model? In the context of cosmology, we may describe it as a simple, idealized description of the universe which reproduces its broad features as observed by astronomers: in particular, the Hubble phenomenon. As we have elected to work within the framework of general relativity, we have to face one difficulty right away: the equations of general relativity are extremely hard to solve. To get exact solutions like the one obtained by Schwarzschild, we have to make simplifying assumptions.

However, in making these assumptions for our cosmological model, we have to go to the opposite extreme to Schwarzschild. Instead of considering an isolated mass in an otherwise empty space, we must consider a space uniformly filled with masses, the galaxies.

In Chapter 1 we saw that the universe appears to be full of galaxies, right to the limit of observations. The galaxies are distributed largely in clusters, which in turn appear to be grouped in larger structures, the superclusters. The space between galaxies is either empty, or it contains matter in invisible form. Thus, strictly speaking, the distribution of matter in the universe is not homogeneous on a galactic scale.

But here we make our first compromise with reality in order to be able to solve Einstein's equations. We assume that the distribution of matter in the universe is homogeneous on a sufficiently large scale—on the order of 1–10 Mpc, say. An analogy which may help us see the reasonableness of this assumption is a fluid which fills a given volume. On the large scale the fluid distribution may appear to be uniform; but on the fine scale we know that each fluid element is made up of molecules moving at random, with voids in between. Likewise, in cosmology, the overall size of the observable universe, 1,000–3,000 Mpc, is so large that inhomogeneities on the scales of galaxies and clusters may be ignored in a simplified model.

Further, we know from the observations which led to Hubble's law that there is nothing special about our location in the universe. Had we observed the universe from another galaxy, we would still have discovered Hubble's law. Thus the large-scale structure of the universe does appear to be homogeneous.

Indeed, as we saw in Chapter 1, the universe is not just homogeneous; it is also isotropic—that is, our observations do not single out any privileged direction. Such a direction could have emerged, for example, if the universe were rotating about an axis, or if it possessed a higher concentration of galaxies in a certain direction. No such privileged direction is indicated by astronomical observations so far.

Thus we will assume, as a starting-point, that the distribution of matter in the universe is homogeneous and isotropic, and *is always so*. The italics emphasize that this is an assumption, an extrapolation into the past and the future, based on the present behaviour of the universe. This assumption is known as the 'cosmological principle'.

The cosmological principle simplifies our search for a model considerably since the homogeneity and isotropy of matter in the universe will be reflected in the geometrical properties of the space occupied by that matter. Hence, the space itself must be homogeneous and isotropic. Mathematics tell us that there are not many such spaces. And those which exist are all characterized by the property of having constant curvature, which of course could be zero, positive, or negative. Mathematicians H. P. Robertson and A. G. Walker independently arrived at the above simplified picture in 1935–6 through rigorous arguments. Hence such space–times are called 'Robertson–Walker space–times'.

Earlier we encountered examples of two-dimensional spaces with zero, positive, and negative curvature. A flat table-top has zero curvature. The surface of a spherical ball has positive curvature which is everywhere the same. A rectangular hyperboloid of revolution has constant negative curvature. These examples are illustrated in Fig. 2.7. Intuition backed by mathematics enables us to visualize their counterparts in three dimensions.

We therefore need only two bits of information to specify the geometry of space: the magnitude and sign of its curvature. The latter is usually expressed by a parameter k, which can take the values 0, 1, or -1. Case I of $k = 0$ describes flat (Euclidean) space. Case II with $k = 1$ is often called 'closed' space, because of its compact and finite volume (compare this with the surface of a sphere). Case III with $k = -1$ is referred to as 'open' space. In cases I and III the volume of the space is infinite.

The magnitude of the curvature is best expressed with the

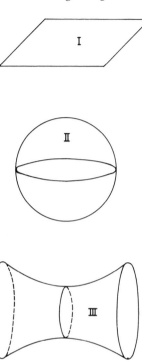

Fig. 2.7. The three types of geometries shown here have their analogues for three-dimensional space too. They describe homogeneous and isotropic universes.

aid of the so-called scale factor, which we will denote by S. The scale factor simply denotes the extent of linear magnification in the separation between two neighbouring points of the expanding space. The statement that the universe is expanding is thus expressed by the mathematical property that S increases with time. The actual curvature of space is given by k/S^2. In cases II and III, therefore, the magnitude of the curvature decreases as the universe expands.

The alert reader may have begun to wonder about the notion of time which has crept into our discussion. Didn't special relativity get rid of the Newtonian idea of an absolute and universal time? Didn't general relativity carry the notion of an 'observer-dependent time' even further along the road? Why, then, are we talking about a single time which applies to the large-scale behaviour of the universe?

The reason that we are able to make such a statement as '*S* increases with time' is that we have imposed a large-scale symmetry on the structure of the universe through the cosmological principle. Recall that the property of isotropy of the universe was based on our observation that the universe looks the same in all directions. This property is not observed by *all* observers at any given point in space. Suppose, for example, we decide to observe the universe not from our position of rest on the earth but from a rocket moving with a speed close to that of light. We would then see some of the galaxies (mainly those lying in our path) as blue-shifted because of the Doppler effect (see Chapter 1). Likewise the galaxies behind us would be more strongly red-shifted than those at right angles to our direction of motion. This situation is illustrated in Fig. 2.8.

There is thus said to be a *unique* observer at each point in space, to whom the universe looks isotropic. Such an observer is called a 'fundamental observer'. As a first approximation to the idealized fundamental observer, we have taken above an earth-based observer. A better approximation would be a sun-based observer; and an even better one an observer situated at the centre of our galaxy. Is the galactic centre the best approximation to the location of a fundamental observer? Not quite! Galaxies also have random motions in clusters. But for the time being we will assume that galaxies as a rule are good approximations to the best location for fundamental observers.

Imagine therefore a typical fundamental observer using a clock to time cosmic events. By the cosmological principle, all fundamental observers will observe exactly similar large-scale behaviour in the universe. It is therefore immaterial whose clock we adopt for timing changes in the large-scale structure of the universe. The time kept by *any* fundamental observer is called 'cosmic time'. It is with respect to this time, t, that the scale factor S is said to increase.

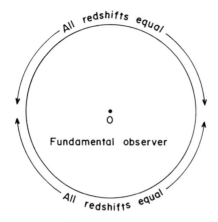

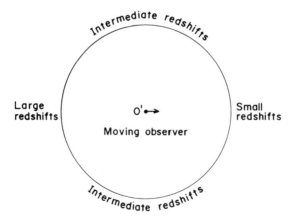

Fig. 2.8. To the fundamental observer, O, the universe looks isotropic. To any other observer, O', moving relative to O, the universe
looks asymmetric.

Going back to our rocket example, we see that the large-scale isotropy of the universe does provide us with a preferred frame of reference. Any observer moving relative to the fundamental observer would find the isotropy destroyed. The preferred frame in which the fundamental observer is at rest is called the 'cosmological rest frame'.

This emergence of a special frame of reference is a manifestation of 'broken symmetry'. General relativity is formulated in such a way that all observers can use their physics with equations having the same formal structure. There is no preferred observer: all are equivalent. The cosmological principle breaks this symmetry, however, and allows us to identify a special set of fundamental observers for whom the universe looks especially simple. This breakdown of symmetry does not violate general relativity, which can still be used to work out how the universe looks to a non-fundamental observer.

Hubble's Law Revisited

Before proceeding further with relativistic cosmology, let us look back at the observational result which started it all. How does the expanding universe picture developed so far lead us to Hubble's law?

The notion of expansion enters through the time-dependent scale factor S. Suppose, as shown in Fig. 2.9, a fundamental observer, O, looks at a galaxy, G, at time t_0. Because light takes a finite time to travel from G to O, the light waves from G must have set off at an earlier epoch t_1 in order to arrive at O at time t_0. Let us suppose that S has been an increasing function of t all the way from t_1 to t_0, so that $S_1 \equiv S(t_1) < S(t_0) \equiv S_0$.

How do light waves travel in curved space–time of the Robertson–Walker kind? We will not go into the mathematical solution to this problem, but will state the result which emerges from it. Suppose a fundamental observer on G measures the wavelength of the outgoing light wave to be λ_1, while

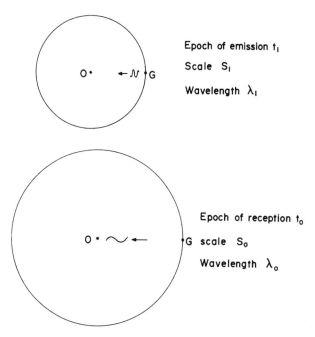

Epoch of emission t_1

Scale S_1

Wavelength λ_1

Epoch of reception t_0

scale S_0

Wavelength λ_0

Fig. 2.9. The scale of the universe grows from S_1 to S_0 during the interval that light from galaxy G travels to observer O. The wavelength of light increases during transit by the factor S_0/S_1.

the observer O finds the wavelength of the incoming light wave to be λ_0. Then the above solution tells us that

$$\frac{\lambda_0}{S_0} = \frac{\lambda_1}{S_1}.$$

That is, both observers arrive at the same answer if they scale down the observed wavelength by their respective scale factors. In the expanding universe, therefore, we will have $\lambda_0 > \lambda$. The red-shift observed by Hubble is given by

$$z = \frac{\lambda_0 - \lambda_1}{\lambda_1} = \frac{S_0}{S_1} - 1.$$

The farther G is from O, the longer the time interval $t_0 - t_1$, and the larger the ratio S_0/S_1; which is why the red-shift is found to increase with distance.

To obtain Hubble's velocity–distance relation for nearby galaxies, let us suppose that G is not very far away from O. We can therefore use the rate of change of S at t_0 to estimate the value of S at t_1:

$$S(t_1) \approx S(t_0) - \dot{S}(t_0)(t_0 - t_1),$$

where $\dot{S}(t)$ denotes the temporal rate of change, dS/dt evaluated at time t. We therefore find that in this approximation

$$z \approx \frac{\dot{S}(t_0)}{S_0}(t_0 - t_1).$$

However, we may compute the distance of the galaxy, D, from the time the light has taken to traverse it; thus

$$D = c(t_0 - t_1).$$

Putting the above two relations together, we get Hubble's law:

$$z = \left(\frac{H}{c}\right)D,$$

provided we identify Hubble's constant, H, with the quantity $\dfrac{\dot{S}(t_0)}{S_0}$.

Friedmann Models

The unknown quantities of space–time geometry having been reduced to just two, k and S, our next job is to determine them using Einstein's equations. To do this, we need information about the matter–energy content of the universe.

We have already made some progress in this direction, since we saw in the last section that the contents of the universe can be compared to a fluid. Now a fluid is characterized mainly by its density, ρ, and its pressure, p. It has other properties too, like viscosity, conductivity, and so on; but we will ignore these for our cosmological fluid, since they are unimportant, at least in the present state of the universe. What about temperature? The temperature, T, can be determined from the pressure and the density if we know more about the physical behaviour of the fluid. We will discuss the behaviour of temperature in Chapter 3.

The Einstein equations, normally ten in number, reduce to just two in our case, because of the simplifying symmetries imposed on the problem. These equations are:

$$\frac{\dot{S}^2 + kc^2}{S^2} = \frac{8\pi G\rho}{3},$$

and

$$2\frac{\ddot{S}}{S} + \frac{\dot{S}^2 + kc^2}{S^2} = -\frac{8\pi Gp}{c^2},$$

where $\dot{S}$ and $\ddot{S}$ are the first and second derivatives of the scale factor, S, with respect to time, t. The mathematical model of the expanding universe will emerge from these equations.

But, first a flashback to history! It was Einstein himself who, in 1917, made the first attempt to construct a relativistic cosmological model. Although in 1917 Slipher's observations of nebular red-shifts were known, their significance and universality had not been established. Einstein, in common with astronomers of his time, assumed, therefore, that the universe in the large is static. But this assumption ran him into difficulty. The above two equations explain why. If the universe is static, S cannot depend on t. Setting $\dot{S} = 0$ and $\ddot{S} = 0$ in these equations, we get

$$\frac{kc^2}{S^2} = \frac{8\pi G\rho}{3} = -\frac{8\pi Gp}{c^2}.$$

For the density, ρ, to be positive requires $k = +1$. But then we get a *negative* value for pressure which is very large, being equal to $-\rho c^2/3$, which clearly doesn't correspond to anything in the physical world.

Faced with this dilemma, Einstein sought to modify his equations of gravity; for it was his firm belief that general relativity should provide not only *a* solution to the cosmological problem, but that the solution should be unique. It was his hope that the modified equations would tell us how the large-scale geometry of the universe is uniquely determined by its matter content. The modification, known commonly as the 'λ-term', introduced a universal force of repulsion, the force of repulsion between any two particles separated by a distance r being λr. Of course, in Einstein's equations, the λ-term does not appear explicitly as a force, but, like gravity, is incorporated in the space–time geometry. Its effect on the above two equations is to change them to

$$\frac{\dot{S}^2 + kc^2}{S^2} - \frac{1}{3}\lambda = \frac{8\pi G\rho}{3},$$

and

$$\frac{2\ddot{S}}{S} + \frac{\dot{S}^2 + kc^2}{S^2} - \lambda = -\frac{8\pi Gp}{c^2}.$$

Since there is no large-scale random motion of galaxies, Einstein felt justified in setting $p = 0$. He thus arrived at a viable model with $k = +1$, $S = S_0 = $ a constant, and $\rho = \rho_0 = $ a constant. Hence

$$S_0^2 = \frac{1}{\lambda}, \quad \text{and} \quad \rho_0 = \sqrt{\left(\frac{\lambda}{4\pi G}\right)}.$$

In this so-called *Einstein universe*, ρ_0 is related through λ to the quantity S_0 which characterizes the linear dimensions of the universe. The Einstein universe has a finite volume, given by $2\pi^2 S_0^3$.

This 1917 model appeared to realize Einstein's hope of a unique and realistic model of the universe. The matter content of the universe determined the geometry of space–time uniquely, and the so-called cosmological constant, λ, was small enough not to interfere seriously with the Newtonian law of gravity so well tested in the solar system.

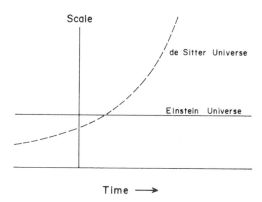

Fig. 2.10. The Einstein model (S = constant) is shown by the un-broken, horizontal line, while the de Sitter model (S = exp Ht) is shown by the dashed line.

In the ensuing years the model fell apart on both scores. Shortly after Einstein proposed his model, also in 1917, W. de Sitter published another solution of Einstein's modified equations:

$$k = 0, \rho = 0, p = 0, S = \exp Ht, 3H^2 = \lambda.$$

The *de Sitter universe* is empty, but test particles in it would recede from each other in accordance with the exponential function. Eddington contrasted the de Sitter universe with the Einstein universe, using the term 'motion without matter' for the former, 'matter without motion' for the latter. Although unrealistic in being empty, the de Sitter universe was prophetic in anticipating the expansion of the universe. Fig. 2.10 illustrates how the scale factor in the de Sitter model changes with time. The model has played a significant role in cosmology since then in contexts which are entirely different, as we shall see later.

Thus Einstein's model turned out to be neither unique nor realistic in describing the expanding universe. Was the λ-term therefore unnecessary? Einstein thought so in the 1930s, calling it the 'greatest blunder' of his life. For the time being we will bury it, but not very deep, for we will need to resurrect it later!

If we are not going to insist on the universe being static, we might as well go back to the unmodified equations discussed earlier. In 1922 Alexander Friedmann from the USSR solved these equations to obtain simple models of the expanding universe. Coming *before* Hubble's observations of 1929, the Friedmann models were initially ignored as mere theoretical exercises; but subsequently they came to occupy centre stage in cosmology.

As in the Einstein model, we set $p = 0$. (We will look more closely at this assumption in the next chapter.) It is then easy to solve the two equations and to place the solutions in three classes. In all cases the density satisfies the simple relation ρS^3 = a constant. This relation is to be expected in a system which conserves matter; for, as the system expands, its density diminishes in inverse proportion to its volume.

The class I model is unique and is given by $k = 0$. Since $k = 0$ denotes a flat three-dimensional space occupied by galaxies, this model is often called the 'flat model'. Promoted in 1932, in a joint paper by Einstein and de Sitter, the model is also called the 'Einstein–de Sitter model'. The scale factor and density in this model are simple power-law functions of time:

$$S \propto t^{2/3}, \ \rho \propto \frac{1}{t^2}.$$

It is convenient to express the Hubble constant, H, as a function of time also. We have already seen how H is related to S and its rate of change with time, $\dot{S}$. In this case we have

$$H = \frac{\dot{S}}{S} = \frac{2}{3t}.$$

A simple relation exists between H and ρ which is valid for all epochs in this model:

$$\rho = \rho_c = \frac{3H^2}{8\pi G},$$

where ρ_c is the 'closure density'. In Chapter 1 we mentioned the prevailing uncertainty in the measurement of H. Taking the present value of Hubble's constant, $H = 75 \text{ km s}^{-1} \text{ Mpc}^{-1}$, we find that $\rho_c \cong 10^{-29} \text{ g cm}^{-3}$. To understand the significance of ρ_c, we need to look at models of classes II and III.

The equation determining S as a function of t can be solved in the cases $k = \pm 1$, although the resulting solutions are not as simple to express as in the $k = 0$ case. Moreover, two parameters are needed to specify the solution. It is convenient to label a typical member of the family of such solutions by the values of the Hubble constant, H_0, and the density parameter *at the present epoch*, Ω_0. The latter parameter is simply the ratio $\Omega = \rho/\rho_c$ evaluated at the present epoch.

Two differences become immediately noticeable in the two classes of models. For models of class II (with $k = +1$) the density parameter, Ω, is *always* greater than 1, while for models of class III ($k = -1$), Ω *always* lies between 0 and 1. Since the space in class II models is closed, while that in class III models is open, we have here a potential test of cosmological theory, a test that relates a global property of geometry of space to its matter content. If the mean matter density of the universe at the present epoch exceeds ρ_c, we are living in a closed universe ($\Omega_0 > 1$), while if ρ is less than ρ_c ($\Omega_0 < 1$), the universe is open. The unique case I model thus stands poised precariously between the open and closed types. The crucial physical quantity that tips the balance one way or the other is the parameter $\Omega_0 = \rho/\rho_c$, which explains why ρ_c is called the 'closure density'.

The second difference between class II and class III models is shown in Fig. 2.11: it lies in the behaviour of the scale factor, S. In the latter, S keeps on increasing with t, always at a rate faster than that for the flat Einstein–de Sitter model. In the former, on the other hand, S increases with t up to a certain epoch and then begins to decrease again.

As far as past behaviour is concerned, all models agree qualitatively that there was an instant in the past, which we shall denote by $t = 0$, when the scale factor was zero. As far as future behaviour is concerned, there are disagreements, however: models of classes I and III have the universe expanding for ever, whereas models of class II have the

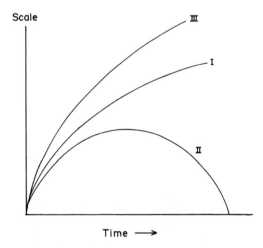

Fig. 2.11. Typical expansion curves for the three classes of Friedmann models. In models of class II the universe first expands and then contracts, whereas in models of classes I and III the universe continually expands.

universe expanding and then recontracting to a state in which S becomes zero again.

The epoch $t = 0$ when S was zero and the density infinite must have been an epoch of violent activity, for the Hubble constant itself was infinite at $t = 0$, implying explosive expansion. Hence this epoch is called the epoch of the 'big bang'. Before we study this important epoch, which is identified with the origin of the universe, we will highlight two important properties of the Friedmann models.

First, it is easy to calculate the age of the Friedmann universe as the time elapsed since the big bang. In fact, we have already done so for the Einstein–de Sitter model, for which we found a simple relationship between the Hubble constant and time. The age of this model universe turns out to be

$$t_0 = \frac{2}{3H_0} \cong 9 \times 10^9 \text{ years,}$$

taking the currently accepted value of H_0, 75 km s^{-1} Mpc^{-1}.

Simple integral calculus is needed to compute t_0 for class II and III models. For the mathematically oriented we give the answer below:

For class II ($\Omega_0 > 1$):

$$t_0 = \frac{1}{H_0}\left\{\frac{\Omega_0}{2(\Omega_0 - 1)^{3/2}}\left(\sin^{-1}\frac{\Omega_0 - 2}{\Omega_0} + \frac{\pi}{2}\right) - \frac{1}{\Omega_0 - 1}\right\}.$$

For class III ($\Omega_0 < 1$):

$$t_0 = \frac{1}{H_0}\left\{\frac{1}{1 - \Omega_0} - \frac{\Omega_0}{2(1 - \Omega_0)^{3/2}}\ell n\left(\frac{2 - \Omega_0}{\Omega_0} + \frac{2(1 - \Omega_0)^{1/2}}{\Omega_0}\right)\right\}.$$

It is found that the age of the universe decreases as Ω_0 increases. So class II models have ages less than $2/3H_0$, whereas class III models have ages which are greater, the maximum possible age being $1/H_0 \cong 13$ billion years for the extreme case of an empty universe ($\Omega_0 = 0$)! Fig. 2.12 illustrates how the age of the universe decreases as Ω_0 increases.

It is here that we recall from Chapter 1 the estimate of 4.32 billion years for the age of Brahma according to ancient Hindu thinkers. As far as I know, of all the various estimates contained in religious or ancient philosophical writings, this comes closest to the modern time-scales associated with the universe.

The second important point concerns the temporal behaviour of the parameter Ω. Although at the present epoch, t_0, Ω_0 for a class II or a class III model may differ significantly from its value $\Omega_0 = 1$ for the flat model, the difference, ($\Omega - 1$), was smaller in the past. Indeed, the closer we approach the big bang epoch, the smaller was this difference. In other words: whatever initial conditions in the early history of the universe determined its subsequent behaviour would have had to be very finely tuned to give a model moderately *different* from the flat model at the present epoch. We will elaborate this point further in our discussions of the early history of the universe.

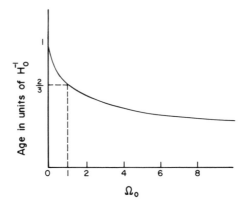

Fig. 2.12. The age of the universe according to Friedmann models with different Ω_0, expressed in units of H_0^{-1}. The age $t_0 = 2/3H_0$ corresponds to the class I (Einstein–de Sitter model). Note that the absolute upper limit for the age is H_0^{-1}.

The Big Bang

Our simplified mathematical solution of the cosmological problem has thus led us to a profound conclusion: that the universe began under somewhat extreme conditions about 10 billion years ago (the exact value depending of course on H_0 and Ω_0). The reason for identifying the big bang epoch with a 'beginning' is more mathematical than physical, and it derives from a weakness in the present formulation of general relativity.

If we take the situation $S = 0$ seriously, it implies zero volume for the space of all fundamental observers. It also implies that several of the parameters characterizing the non-Euclidean geometry of space–time become infinite. Mathematically, therefore, we cannot talk meaningfully of this 'state' of the universe; nor can we probe beyond it and ask what the universe was like *before* $t = 0$. Since the physicist requires a well-behaved space–time to study the operation of any natural processes, he is therefore forced to confine his attention to the period $t > 0$.

Most physicists would agree that the singular situation at $t = 0$ reflects an incomplete understanding of how gravity operates when matter is in an extremely dense state. General relativity cannot therefore be relied on to describe the universe at $t = 0$. Opinions differ, however, as to what the 'correct picture' might be. We will return to this question and to some of the alternatives available in Chapters 6 and 7.

Taking a pragmatic view, let us ask whether we have any *observational* evidence that the universe originated in a big bang. First, we note right away that there is no direct evidence in the probes made hitherto of remote parts of the universe. Recall that observations of galaxies with large red-shifts tell us about the earlier epochs of the universe. The largest red-shifts observed for galaxies are of the order of $z = 1$, which, using our red-shift formula (see p. 57), takes us to epochs when the scale factor was half its present value. Even assuming that the so-called quasi-stellar objects are very far away, and that their red-shifts obey Hubble's law, we go only as far back as the time when $S_1 \cong S_0/5$ (the largest z-value to date is 4).

All this is a far cry from the $S = 0$ epoch which we are trying to reach. We therefore have to rely on other, indirect evidence. Here we are like archaeologists probing ancient history through the study of relics. Are there any relics which astronomers can point to as proofs that the universe did indeed have a violent beginning? We will take up this question in the next chapter.

3

Relics of the Big Bang

The telescope at one end of his beat,
And at the other end the microscope,
Two instruments of nearly equal hope.

Robert Frost

If a cosmologist today is asked to identify the two observational results that have had the most profound impact on the development of twentieth-century cosmology, he will of course point to Hubble's law. Which is the other?

The second result was made known in 1965, in the *Astrophysical Journal*, under the decidedly unpretentious title 'A Measurement of Excess Antenna Temperature at 4080 Mc/s'. Its authors, Arno Penzias and Robert Wilson, two scientists at the Bell Telephone Laboratories at Holmdale, New Jersey, had found the 'excess' radiation entirely serendipitously in the following sense. They were engaged in an experiment to detect radio waves coming from the disc of the Galaxy. To this end, they were testing their horn-shaped antenna for unwanted radio noise which might contaminate their measurements. Expecting less of a radio contribution from the Galaxy at the high microwave frequency of 4,080 megahertz (wavelength 7.35 cm), Penzias and Wilson had tuned their antenna to this frequency. Although they were looking for sources of contamination, it came as a surprise to them when their antenna started picking up radio noise *uniformly from all directions*.

This was in the spring of 1964. The lack of directionality of the radiation precluded any particular source as the cause. Penzias and Wilson carefully ruled out any instrumental effects, large or small (including the coating of the antenna by pigeon droppings!). What could this noise be due to?

The answer came eventually, not from radio engineers, but

from cosmologists. Robert Dicke and Jim Peebles at Princeton had been interested in studies of the 'early universe', of what went on during the first few minutes after the big bang. To them the Penzias–Wilson discovery came as no surprise; their work had already led them to expect such a radiation background as a relic of that early epoch of the universe. This discovery of microwave background radiation led to a revival of interest in the big bang universe. Indeed, as cosmologists began turning the pages of recent history, they came across the pioneering work of George Gamow and his colleagues Ralph Alpher and Robert Herman in the late 1940s. It was theoretical work on the early universe which in many ways anticipated the Penzias–Wilson discovery.

But how can an apparently sourceless background of microwave radiation existing now have anything to do with what went on in the universe billions of years ago? The answer to this question lies in one of the most daring applications of physical ideas to cosmology—daring, because the extrapolation of the physical laws involved takes us well beyond the limits within which they have been verified in terrestrial laboratories.

In this chapter we will concentrate on this early phase in the life of the universe, when it was barely a second or so old. The picture we will work with is vastly different from the one we encountered in Chapter 2. Yet, to work back to the early history of the universe, we must start from its present state.

Matter versus Radiation

In our description of cosmological models in Chapter 2, we made one simplification. We treated cosmic matter as a pressure-free fluid. Let us examine the extent to which this simplification is valid in the universe today, and whether it needs to be modified in considering the early universe.

Kinetic theory of gases tells us that pressure in a fluid arises from random motions of its atoms and molecules. When a fluid 'flows', it has an overall motion in the direction of flow. However, if we were to examine a typical fluid element carefully, we would find that, superposed on the bulk motion, are microscopic movements of its atoms and molecules, at varying speeds and in varying directions. These random motions average out to zero net speed; but they have the overall effect of generating the property of pressure in a fluid. As a result of this property, any surface held inside the fluid feels a force in a direction perpendicular to its area.

In the cosmic fluid we have to enlarge this picture from atoms and molecules to galaxies. We have already seen that the cosmic fluid partakes of the bulk motion of expansion. Superposed on this motion are the random movements of galaxies in clusters, which have speeds on the order of 300 km s^{-1}, that is, around a thousandth of the speed of light. The dynamical effect of these speeds on the expansion of the universe is measured by the square of this velocity ratio:

$$\frac{p}{\rho} \sim \left(\frac{V}{c}\right)^2 \sim 10^{-6}.$$

In other words, if we take note of the pressure of the cosmic fluid arising from random motions of galaxies, we find that its effect on the rate of expansion is negligible compared with the effect of the fluid density which we have already taken into account.

Further, it is the property of such random motions that they tend to die out as the expansion continues. The drop in the speed of a random motion, V, is in inverse proportion to the universal scale factor, S: that is, $V \propto 1/S$.

In other words, as the universe continues to expand, the effect of pressure in the future will be even less than it is now.

By the same token, however, the above formula warns us not to be so complacent when considering the past. For if S were smaller than its present value by a factor of 1,000, we would find V comparable to c, and the pressure no longer negligible! It is believed that if we extrapolate the formula this far back in time, we will inevitably reach epochs when galaxies themselves came into existence. The pressure of cosmic fluid at that time would have arisen, therefore, not from galaxies (as yet unformed), but from pre-galactic material, which was presumably in a more elementary state like hot gas. Thus we are led to a picture of the cosmic fluid in early times as a hot gas, with pressure arising from random motions of its constituent particles. And this particle motion would be expected to be more and more violent the further back in time we push our cosmological considerations.

Let us look at this violent activity from another standpoint. A gas whose component particles are moving so rapidly as to have speeds comparable to the speed of light is similar in its behaviour to pure radiation. Radiation is made up of photons, which naturally always travel at the speed of light. Radiation made of randomly moving photons has pressure much in the same way that a gas has pressure. Moreover, there is a simple formula which connects the pressure of radiation to its energy density, ε:

$$p = \frac{1}{3}\varepsilon.$$

This result holds as the universe continues to expand. Just as expansion of the universe reduces the random motions, and hence the pressure in the cosmic fluid, so it reduces the pressure and the energy density of any radiation present in the universe. The formula relating p and ε to the scale factor, S, is a simple one:

$$p = \frac{1}{3}\varepsilon \propto \frac{1}{S^4}.$$

The pressure and energy density of radiation thus fall off with expansion more rapidly than the density, ρ, of pressure-free cosmic matter. For we saw in Chapter 2 that ρ falls off as the inverse cube of S.

In other words, the balance shifts from radiation to matter in a universe which continues to expand. In the early universe radiation was dominant, whereas in the present universe matter is dominant, as shown in Fig. 3.1. Can we roughly estimate the epoch when the switch-over occurred? We can, if we know the present values of ε and ρ.

The matter density of the universe, ρ, is believed to be made

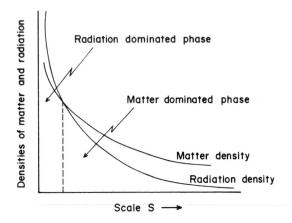

Fig. 3.1. The densities of matter and radiation both fall off as the universe expands. Whereas matter dominates today, there was an earlier era in which radiation dominated. The epoch of transition from the radiation- to the matter-dominated phase is where the two curves intersect.

up of two components, one visible, the other invisible. The visible component is mainly in the form of galaxies, which, as we saw in Chapter 1, occur singly or more often in clusters. This visible component of density is estimated at around 3×10^{-31} g cm^{-3}. The invisible component is anybody's guess; we will have more to say about it in later chapters. Opinions differ as to how much density resides in hidden matter. Estimates going as high as the closure density, $\rho_c \cong 10^{-29}$ g cm^{-3}, or even higher have been quoted by experts in the field. On the other hand, there are those who question the very existence of matter in any form other than the visible. As Table 3.1 summarizes the energy densities of the background radiation of various wavelength regions, ranging from radio waves at long wavelengths to gamma rays at the shortest wavelengths, it is clear immediately that the bulk of the radiant energy in the universe resides in the microwaves, whose existence was first detected by Penzias and Wilson.

Table 3.1 Radiation Backgrounds in the Present Universe

Form of radiation	Wavelength range (cm)	Estimated energy density (erg cm^{-3})
Radio waves	≥ 7.3	$\leq 10^{-18}$
Microwaves	$0.1-80$	$\sim 6 \times 10^{-13}$
Infrared	$10^{-4}-10^{-3}$	$\sim 4 \times 10^{-13}$ (tentative)
Visible	$4 \times 10^{-5}-8 \times 10^{-5}$	$\sim 3.5 \times 10^{-15}$
Ultraviolet	$2 \times 10^{-6}-2 \times 10^{-5}$	$\sim 10^{-16}-10^{-15}$ (tentative)
X-rays	$3 \times 10^{-9}-1.2 \times 10^{-7}$	$\sim 10^{-16}$
Gamma rays	$<1.2 \times 10^{-13}$	$\sim 2 \times 10^{-17}$

Applying the Einstein mass–energy conversion formula $E = mc^2$, it is also clear that the radiation energy density falls short of the matter density, even taking its visible form only, by a large factor:

$$\frac{\varepsilon_0}{\rho_0 c^2} \lesssim 10^{-3}.$$

To find the answer to our question, when in the past was radiation more dominant than matter?, we now recall that the ratio $\varepsilon/\rho c^2$ varies inversely with the scale factor; thus

$$\frac{\varepsilon}{\rho c^2} \propto \frac{1}{S}.$$

This relation tells us that to make up for the shortfall of a factor of 1,000, we must reduce S by the same factor. The scale factor of the universe must therefore have been smaller than its present value by a factor $\sim 1,000$ at the time when the densities of radiation and matter were comparable (see Fig. 3.1).

In Chapter 2, we found in the red-shift a convenient parameter for comparing the size of the universe at a past epoch with its present size. Using the red-shift in the present context, we can say that the universe was radiation-dominated at red-shifts exceeding $\sim 1,000$.

Back in the mid-1940s George Gamow had appreciated this

fact, and had used it to develop the concept of the 'hot big bang'. As the adjective 'hot' implies, Gamow was concerned with the early history of the universe when it was radiation-dominated. At that stage, even particles of matter were moving at speeds comparable to the speed of light, so they were indistinguishable in this regard from particles of radiation, photons. It is now possible to give a quantitative meaning to the word 'hot'. Let us see how it is done.

Thermodynamic Equilibrium and the Black-body Radiation

It was Gamow's thesis that the hot brew of matter and radiation in the early universe was bound to undergo nuclear transformations, in which simpler constituents of matter combined to form the bigger units which are the various atomic nuclei found in the universe today. But before discussing Gamow's ideas, let us take a look at thermodynamics, which deals with physical processes involving exchanges of heat with other forms of energy.

Thermodynamics brings to physics two new notions: temperature and 'entropy'. (Entropy we will discuss in a later chapter.) To understand the notion of temperature, consider the following experiment. We have two gases in separate containers, A and B, say. The constituent particles of the gases are moving at random, although the gases themselves are at rest in their respective containers. Suppose the overall activity of particles of gas in A is more sluggish than that of particles in B.

What happens when A and B are connected by a channel? The particles can then flow freely from one container to the other. They can also collide, and in this process the more energetic particles will share their energy with their less energetic neighbours. Soon the more sluggish particles of gas originally in A begin to move more energetically, at the expense of their initially more active neighbours from B. The give and take continues between individual pairs of particles as they collide and disperse, but the situation is soon reached in

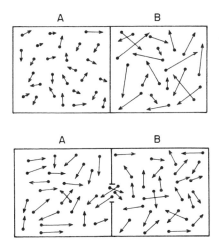

Fig. 3.2. The arrows attached to the points representing the gas particles are broadly indicative of their velocities. In container B there is more activity than in container A. If the particles in the two containers are allowed to mix, they will finally settle down to a state of activity midway between the two initial states.

which the average energy of random motion per particle is the same for particles originating in A and B (see Fig. 3.2).

This situation is called the state of 'thermodynamic equilibrium'. When it is reached, we can ascribe a 'temperature' to the system as a measure of the average energy per particle. The larger this energy, the higher the temperature. It is found that if a system not in thermodynamic equilibrium initially is left to itself, it will reach a state of equilibrium as long as opportunity is provided for individual constituents to mix and share their energies.

The subject of *statistical mechanics* goes deeper into the analysis of such equilibrium situations. For example, it can tell us the momentum distribution of particles in a gas which has attained thermodynamic equilibrium. Although it is the average kinetic energy per particle which quantifies the thermodynamic property of temperature of a gas, equilibrium does *not* mean that all gas particles have the same energy of motion. Just as a population of human beings subject to economic forces in a free society has its distribution of rich and poor, so a gas in thermodynamic equilibrium has a distribution of particles with high and low momenta in specified proportions.

As the adjective 'statistical' indicates, such distribution functions tell us how large populations of particles behave, rather than how any individual particle moves or collides. With regard to the early universe, we need to make two kinds of distinctions among particles. The first is whether the particle is a 'fermion' or a 'boson'. This distinction is a legacy of quantum theory, which tells us how to treat a system of identical particles. We will save the details for Chapter 4; here we will simply state what the distinction implies. If the state of a particle in a system is specified by a certain set of parameters—say, energy, angular momentum, spin, and so on—then the rule is that no two fermions can ever occupy the same state. Bosons, on the other hand, are not subject to the same restriction.

This difference, coupled with the fact that, unlike particles in classical mechanics, neither fermions nor bosons retain their individuality during their motion and interaction with one another, leads to new types of statistics for handling large populations. These statistics are the so-called Fermi–Dirac statistics for fermions and Bose–Einstein statistics for bosons. (The names are those of the scientists who first studied these statistics.) Accordingly, we get different distribution functions for families of fermions and bosons. As a rule, particles of half-odd integral spin ($\frac{1}{2}$, $\frac{3}{2}$, ...), like electrons, protons, neutrinos, and so forth, are fermions, whereas particles of zero or integral spin (o, 1, 2, ...), like mesons and photons, and bosons.

The second distinction is that between relativistic and non-relativistic particles. To understand this distinction, suppose a particle has mass m_0 when it is at rest. According to the theory of special relativity, when this particle is moving with speed v, it will have an energy, E, given by

$$E = \frac{m_0 c^2}{\sqrt{\left(1 - \frac{v^2}{c^2}\right)}}.$$

It is the average value of E per particle which tells us the equilibrium temperature of the population. We express the average of E by $\langle E \rangle$, and write the temperature, T, as

$$T = \frac{1}{k} \langle E \rangle,$$

where k is 'Boltzmann's constant' and is used to convert energy measured in ergs to temperature measured in degrees Kelvin (K). (Zero on this temperature scale is $-273°C$, and a step of $1K$ equals $1°C$.) The constant k in these units is 1.38×10^{-16} erg/K.

Using this constant, we can define a 'rest-mass temperature' for the particle by

$$T_0 = \frac{m_0 c^2}{k}.$$

This temperature is only of theoretical interest. For a given population, a comparison of T with T_0 tells us how energetic or fast-moving the particles in the distribution are. If T is much greater than T_0—that is, if the actual temperature exceeds the rest-mass temperature by a large factor (say ten or more), then we know that the bulk of the particles in the distribution are moving relativistically, with speeds v very close to the speed of light, c. If T is comparable to T_0, then most of the particles are moving at speeds much less than the speed of light, and can be treated non-relativistically.

Table 3.2 gives the values of T_0 for a number of subatomic particles. In this table the rest-mass energy is expressed not in ergs, but in mega-electron-volts, a unit commonly used in nuclear and particle physics (1 MeV $= 10^6$ eV $= 1.6 \times 10^{-6}$ ergs). Notice that the table lists $T_0 = 0$ for photons and neutrinos. At present it is not established whether neutrinos have a non-zero rest mass; we will return to so-called massive

Table 3.2 Rest-Mass Temperatures for Subatomic Particles

Particle	Rest-mass energy (MeV)	Rest-mass temperature (°K)
Photon	0	0
Neutrino	0 (?)	0 (?)
Electron and positron	0.51	5.93×10^9
Muon	105.7	1.22×10^{12}
Pion	135–140	1.6×10^{12}
Proton	938.28	10^{13}
Neutron	939.57	10^{13}

neutrinos later. The present mass limits do suggest, however, that T_0 for neutrinos must be very small compared to T_0 for electrons, the next smallest value in the table.

For photons the equilibrium distribution has a particularly simple form. The momentum and energy of a photon of frequency v are respectively hv/c and hv, where h is Planck's constant. The number of photons per unit volume, n_γ, in an equilibrium distribution at temperature T lying in the frequency band v to $v + \Delta v$ is given by

$$n_\gamma(v)\Delta v = \frac{8\pi}{c^3} \frac{v^2 \Delta v}{\exp (hv/kT) - 1},$$

where γ is the symbol for photons.

If we sum over all frequencies, we get the total number of photons in the distribution per unit volume, N_γ:

$$N_\gamma = 19.2\pi \left(\frac{kT}{ch}\right)^3.$$

Likewise, multiplying the distribution function by the photon energy hv and then summing over all values of v gives us the total energy per unit volume of these photons:

$$\varepsilon = \frac{8\pi^5}{15} \frac{k^4 T^4}{h^3 c^3} \equiv aT^4,$$

where a is called the 'radiation constant', and has the value 7.56×10^{-15} erg cm^{-3} K^{-4}.

The above multiplication of the distribution function by hv gives the intensity distribution of the radiation in thermodynamic equilibrium. This distribution is called the 'blackbody distribution'. The term 'black body' expresses what we have tacitly assumed throughout our discussion of equilibrium distributions so far—namely, that there is no escape or input of energy. All the collisions and the sharing and re-sharing of energy between colliding particles are supposed to occur in a perfectly sealed enclosure. The nearest practical example to a black body is a good-quality oven. When the oven is heated to any specific temperature, the photons of radiation inside it bounce off the walls of the oven, but do not escape from it. Of course, no oven is ideal; there is always *some* loss of energy from its walls, which become warm outside, if not as hot as the interior.

The intensity distribution of a black body always has a peak at a frequency v_0 which increases with its temperature, T:

$$v_0 = 6 \times 10^{10} \ T \ \text{Hz} \ \text{K}^{-1}.$$

(where Hz is short for hertz, which equals 1 cycle per second). This law was discovered experimentally by W. Wien in 1896, but its significance was not realized until the quantum nature of radiation was discovered. This fact is indicated by the appearance of Planck's constant, h, in the intensity distribution, and also in the constant of proportionality (given numerically above) in Wien's law. Classical discussions of thermodynamics as given by Lord Rayleigh and Sir James Jeans led only to an intensity distribution of the form

$$I(v)\Delta v = \frac{8\pi v^2}{c^3} kT \Delta v.$$

That is, there was no ceiling on $I(v)$ in pre-quantum physics. The Rayleigh–Jeans formula is now seen to describe only the low frequency portion ($v \ll v_0$) of the actual black-body distribution. Typical cases of intensity distributions in black bodies at different temperatures are shown in Fig. 3.3.

Let us now go back to the microwave background discovered in 1965 by Penzias and Wilson. Their measurement of

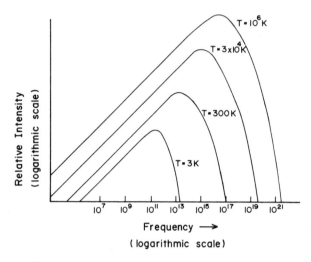

Fig. 3.3. Curves showing the frequency distribution of radiation for black bodies of different temperatures. Notice how the peak of the curve shifts to higher frequencies as the temperature is increased.

the radiation intensity at $\nu = 4{,}080$ MHz lies in the Rayleigh–Jeans region of the black-body distribution. Subsequent measurements—and there have been numerous observations at different frequencies by now—have shown a black-body distribution with a temperature of around 3K. Using Wien's law, the peak intensity of the radiation is found to be at microwave frequencies $\sim 1.8 \times 10^{11}$ Hz. Measurements of the intensity at the peak of the distribution and of the way it drops off beyond ν_0 (that is, for $\nu > \nu_0$) were first made systematically in 1979 by D. P. Woody and P. L. Richards. Since the earth's atmosphere absorbs radiation at these frequencies, measurements had to be made with a balloon-borne interferometer above the main absorbing layers of the atmosphere.

Before we leave the topic of equilibrium distributions, we will mention briefly how the bulk properties of distributions of relativistic particles in general compare with those of the black-body radiation, since we will need this information for our discussions of the early universe, not only in this chapter, but also in chapters to follow.

For relativistic distributions, $T \gg T_0$, and we can ignore the effect of particle rest masses. For a comparison with the photon distribution, we need to know whether our particles belong to the species of fermions or bosons. We also need to know how many different states of spin they have. A particle of spin zero has one degree of freedom, a particle of spin $1/2$ two. The photon, being a particle of spin 1 but zero rest mass, has two spin degrees of freedom. In terms of the black-body radiation of the same temperature (expressed with suffix γ) the particle number density, N, and the energy density ε, are expressed as follows:

For bosons with g_b spin states:

$$N_b = \frac{1}{2} g_b N_\gamma, \quad \varepsilon_b = \frac{1}{2} g_b \varepsilon_\gamma.$$

For fermions with g_f spin states:

$$N_f = \frac{3}{8} g_f N_\gamma, \quad \varepsilon_f = \frac{7}{16} g_f \varepsilon_\gamma.$$

What form does the equilibrium distribution take when the particles of the population are moving non-relativistically? For $T < T_0$, the number of particles of rest mass m per unit volume is given by:

$$N = \frac{g}{h^3} (2\pi m k T)^{3/2} \exp\left(-\frac{T_0}{T}\right),$$

where g is the number of spin states of the particle. Fig. 3.4 illustrates how N changes with T. Notice that because of the exponential function the numbers decline rapidly as the temperature falls. This result has enormous implications for the early universe.

Thermodynamics and Dynamics of the Early Universe

In this chapter we will look at the phase of the early universe originally discussed by Gamow, Alpher, and Hermann.[1] These

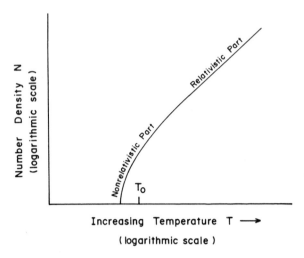

Fig. 3.4. The figure shows how N increases with T for large values of T, but drops sharply as T is lowered below the rest-mass temperature, T_0.

authors assumed that, somehow, soon after the big bang, leptons like electrons, neutrinos, and their antiparticles, as well as baryons like neutrons and protons, were in existence along with radiation. The different particles were in thermodynamic equilibrium at a high temperature. The temperature fell, however, as the universe expanded; but at a certain stage lasting about 3 minutes, conditions were ripe to induce nuclear fusion of neutrons and protons into light nuclei like deuterium, helium, lithium, and so on. In the remainder of this chapter we will develop this picture quantitatively, including inputs that have come since Gamow's pioneering work of the 1940s.

First let us see what form the expansion of the early universe takes in the radiation-dominated phase. Assuming thermodynamic equilibrium at a temperature T, we first note that the predominant contribution to the energy density, ε, and the pressure, p, will come from relativistic particles, since non-relativistic particles will be too few to influence the dynamics of the expansion. For such particles, which naturally include photons, our previous discussion gives a simple formula for ε and p:

$$\varepsilon = 3p = \frac{1}{2}gaT^4,$$

where g is the total *effective* number of spin degrees of freedom for all relativistic species (where $T \gg T_0$). This is defined in terms of the actual spin degrees of freedom of bosons and fermions, g_b and g_f, thus:

$$g = g_b + \frac{7}{8}g_f.$$

The factor 7/8 arises because of the difference between the expressions for ε_b and ε_f for the distributions of fermions and bosons that we obtained earlier.

It is now a simple matter to substitute the expressions for ε and p in the Einstein equations for the homogeneous isotropic models discussed in Chapter 2. The two equations are

$$3\frac{\dot{S}^2 + kc^2}{S^2} = \frac{4\pi G}{c^2}gaT^4$$

and

$$2\frac{\ddot{S}}{S} + \frac{\dot{S}^2 + kc^2}{S^2} = \frac{8\pi G}{3c^2}gaT^4.$$

We will anticipate here that the 'curvature term' kc^2/S^2 on the left-hand side of these equations is unimportant at this stage, and ignore it. In Chapter 5 we will come back to this important point. With $k = 0$, these equations admit a very simple solution: namely,

$$S \propto t^{1/2}.$$

Although the constant of proportionality is not determined, and can always be absorbed in the scale factor, we get a *unique* relationship between time and temperature:

$$t = \sqrt{\left(\frac{16\pi Gga}{3c^2}\right)} T^2.$$

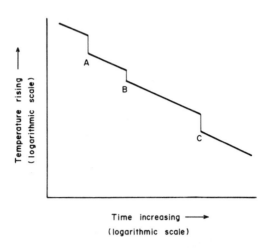

Fig. 3.5. Schematic graph showing the time–temperature relationship in the early universe. The jumps in the curve at points A, B, C, are artificial; they indicate changes in the particulate nature of matter. Normally these changes would occur smoothly, rather than abruptly.

Fig. 3.5 illustrates the time–temperature relationship in the early universe.

The interesting point about this equation is that the constant of proportionality between t and T^2 is completely determined by the physical constants c, G, and a and by the effective number of spin degrees of freedom, g. To fix ideas, suppose T is $\sim 10^{10}$ K. From Table 3.2 we see that apart from photons (which are bosons), all other relativistic particles at this temperature are fermions: electrons, positrons, their neutrinos and antineutrinos, as well as the neutrino–antineutrino pairs of the muon type. (There is also a third species of neutrino, the τ-neutrino; but it is expected to be massive and may be non-relativistic at these temperatures.) For each neutrino and antineutrino we have $g_f = 1$, while for electrons and positrons $g_f = 2$. Our formula therefore gives $g = 2 + \frac{7}{8}(1 + 1 + 1 + 1 + 2 + 2) = 9$. The temperature–time relationship thus becomes $T \cong 1.04 \times 10^{10} \, t^{-1/2}$ K, where t is in seconds. Thus, one second after the big bang the temperature of the universe was just over ten billion degrees!

Before we proceed further, let us explore whether the condition of thermodynamic equilibrium was in fact attainable in the universe at this young age. We have already seen that thermodynamic equilibrium is attained in a gas by frequent collisions of its constituent particles. What exactly are 'collisions'? At the macroscopic level we are accustomed to the picture of colliding billiard-balls, in which a fast-moving ball impinging on a ball at rest imparts to it some of its momentum. If the radius of each billiard-ball is R, then we know that two balls will collide provided their centres are separated by a distance less than $2R$. In other words, the area of the collision zone within which the trajectory of the moving ball must lie in order to collide with the ball at rest is simply

$$\sigma = \pi \times (2R)^2 = 4\pi R^2.$$

This area may be called the 'collision cross-section'. The larger the value of σ, the higher is the probability that two balls moving at random will collide.

At the microscopic level these concepts carry over, but with a few modifications. First, gas particles are not rigid bodies of fixed radius like billiard-balls. Nor are they point-particles moving along definite trajectories as in Newtonian mechanics. In the quantum-mechanical picture a particle is described by a 'wave function', which tells us the probability of finding the particle in a given volume at any given time. In spite of these differences, we can still talk of collisions as events in which, because of the basic interaction between two particles, an exchange of energy and momentum takes place. The collision cross-section in such a case depends of course on how strong the interaction is: the stronger the interaction, the larger the cross-section.

Let us now look at our cosmic brew of electrons, positrons, neutrinos, and photons. What are the interactions through which they 'collide'? The relevant reactions are the electromagnetic and the weak interactions, except that neutrinos are not subject to electromagnetic forces or photons to the weak interaction. From our considerations of collision cross-section, we therefore suspect that the neutrino is only a weakly interacting particle, the one least likely to collide and thereby partake in the process of thermodynamic equilibrium. Let us consider neutrinos first.

The calculations of the electroweak theory briefly referred to in Chapter 2 tell us that the neutrino collision rate with other particles, β, depends on the temperature of the universe:

$$\beta = aT^5 \exp(-b/T),$$

where a and b are constants. Thus, if the universe were static with a temperature T, then, given sufficient time, a population of neutrinos colliding with other particles would reach and subsequently maintain a state of thermodynamic equilibrium.

The universe, however, is not static. The early universe was ·expanding very fast, its rate of expansion being given by the then value of the Hubble constant:

$$H = \frac{\dot{S}}{S} = \frac{1}{2t} \propto T^2.$$

Evidently expansion, by its very tendency to separate any two particles, inhibits collisions and the maintenance of thermodynamic equilibrium. We therefore have to compare β with H in order to decide whether collisions occur frequently enough in the expanding universe.

Fig. 3.6 illustrates how H and β change with temperature. It is clear that at sufficiently high temperatures, β is high enough that thermodynamic equilibrium is achieved and maintained. At lower temperatures, the collision rate cannot keep up with expansion, and the neutrinos cease to have contact with other particles. The critical temperature for this to happen is about 10^{10} K. Thus neutrinos, early on, when the universe was hotter than ten billion degrees, did collide and maintain thermodynamic equilibrium. Subsequently they became isolated and essentially decoupled from the rest of the universe.

What happens to the decoupled neutrinos? Although they no longer interact with other particles, they are still subject to gravity, and hence to the expansion of the universe. Their

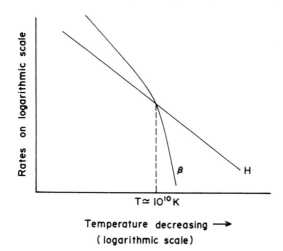

Fig. 3.6. The rate of expansion of the universe, H, and the neutrino collision rate, β, are plotted against the temperature, T, on a logarithmic scale. Notice that at high temperatures $\beta \gg H$, while at low temperatures $\beta \ll H$. The two are comparable near $T = 10^{10}$ K.

number density therefore drops, as $1/S^3$, and their average momentum as $1/S$. If they are really of zero rest mass, their distribution functions will also be reduced with expansion, such that their temperature will fall as $1/S$. If, on the other hand, neutrinos do have a small rest mass, the above conclusion is still valid provided T is greater than T_0. We will return to this point in Chapter 5, and assume for the time being that neutrinos are without rest mass.

By contrast to the weak interaction, the electromagnetic interaction is able to maintain a high enough collision rate even below the temperature of 10^{10} K. However, new considerations enter the picture when the temperature drops to half this value, for a glance at Table 3.2 tells us that the value of T_0 for electrons and positrons is around this value. This is the temperature at which the electron–positron population begins to be non-relativistic. Moreover, during the cooling period from 10^{10} K to 10^9 K, we also have to take note of the tendency of electrons (e^-) and positrons (e^+) to annihilate each other and produce pairs of photons (γ):

$$e^- + e^+ \rightarrow 2\gamma.$$

Of course, if a sufficiently large number of energetic photons is present, the reverse reaction also takes place:

$$\gamma + \gamma \rightarrow e^- + e^+.$$

In thermodynamic equilibrium at temperatures exceeding five billion degrees, both reactions were in fact taking place, thereby maintaining the population of electrons and positrons. As the temperature dropped below this value, however, the number of photons energetic enough to create electron–positron pairs declined sharply, and annihilations began to dominate. By the time the universe had cooled to about a billion degrees, the pairs had been effectively eliminated from the cosmic brew.

This disappearance of the pairs led to an increase in the number of photons. The result was that the photon population came to have a higher temperature than the neutrinos. This increase can be seen by comparison with the temperature of the neutrinos.

Recall that at temperatures exceeding ten billion degrees the neutrinos were as much part of the thermodynamic equilibrium as the photons, and thus had the same temperature. Later, neutrinos ceased to partake in the 'equilibrium through collisions' process, although their temperature continued to decline as $1/S$. Had no fresh photons been injected into the cosmic mix, the temperature of the photon population would also have continued to decline as $1/S$, and would thus have remained equal to the temperature of the neutrino population. The electron–positron annihilation, however, raised the photon temperature (T_γ) above the neutrino temperature (T_ν). This is shown in Fig. 3.7. Exact calculation shows that

$$\frac{T_\gamma^3}{T_\nu^3} = \frac{11}{4};$$

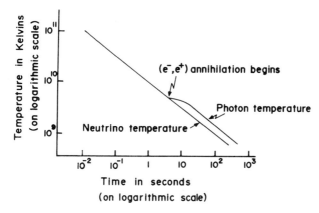

Fig. 3.7. The neutrino and photon temperatures are plotted as functions of time. Notice that shortly after 1 second age of the universe the annihilation of electron–positron pairs raises the photon temperature above the neutrino temperature by a factor ≈ 1.4.

that is, $T_\gamma \approx 1.4\,T_\nu$. Assuming that after this annihilation event nothing happened to alter the numbers of photons and neutrinos in the universe, we would expect this ratio to remain valid to this day. Since the photon temperature as exhibited by the microwave background is around 3 K, the neutrino temperature today would then be expected to be around 2 K. We will discuss the implications of this result when considering the possibility of neutrinos with non-zero rest mass.

Primordial Nucleosynthesis

The stage is now set for the universe to play the role of a thermonuclear reactor in which neutrons and protons at high temperature are brought together to form progressively bigger nuclei of atoms. First let us try to understand why the temperature range $10^9 - 10^8$ K is important for the synthesis of nuclei.

The simplest nucleus, one which needs no synthesis at all, is of course that of hydrogen. It consists of a single proton. The next one on the ladder is the heavy hydrogen, or deuterium, nucleus $^2_1$H, consisting of one proton and one neutron. We will follow the convention of denoting this nucleus, also called the 'deuteron' by the symbol d. Since it contains a neutron (n) and a proton (p) at rest, we would expect its mass, m_d, to be equal to the sum of the masses of the neutron and the proton, m_n and m_p respectively. In practice it is found to be somewhat less than this value, by an amount which we will designate Δm; thus,

$$m_d = m_n + m_p - \Delta m.$$

What is the cause of this deficit? It lies in the strong force which binds the nucleus together. It is the force of attraction which keeps the neutron and the proton together in the small nuclear region of size $\sim 10^{-15}$ m. The strong force has no effect at distances greater than this, so protons and neutrons well separated do not experience it. To strip a deuteron apart, however, we have to do work against this attractive nuclear force. The law of conservation of matter and energy requires that this work should appear as the excess energy which the free neutron and proton have over their bound-state energy. Called the 'binding energy', this excess energy, B, must of course equal Δmc^2. For the deuteron, it is 2.22 MeV.

To appreciate the significance of this figure, we relate the electron-volt to the unit of temperature. At temperature T, the quantity kT denotes the energy; hence 1 eV = 1.16 × 10^4 K. In other words, when we talk about a gas in thermodynamic equilibrium with a temperature of 1.16 × 10^4 K, we know that the energies of typical particles in the gas are ~ 1 eV. Conversely, we may express the temperature of the gas in eV, MeV, or GeV units, as we will frequently do later on. Thus a gas temperature of 1 MeV at once tells us that the average particle energy in it is 1 MeV.

The binding energy of the deuterium nucleus thus translates into an equivalent temperature of $\sim 2.58 \times 10^{10}$ K. This figure suggests that if we have a hot gas of neutrons and protons at a temperature exceeding 2.58 × 10^{10} K, the binding imposed by the strong force is not adequate to hold a deuterium nucleus together. Collisions with fast-moving particles will strip it apart. Evidently we need a cooler temperature than this to form the deuteron. But how cool?

The actual value depends on the numerical distributions of neutrons and protons. So far, we have not considered these explicitly, because there were too few of them to affect the expansion of the universe at the temperatures under consideration. (Recall from Table 3.2 that their T_0 value is 10^{13} K.) We now need to take note of their existence, because they are essential for the formation of nuclei. The first question we have to settle, therefore, is whether the neutrons and protons, being non-relativistic, are able to maintain thermodynamic equilibrium. Because if they are, we can say something definite about their relative abundances at any given temperature. This ratio, as we shall discover shortly, contains crucial, observable information.

It was the Japanese physicist Chushiro Hayashi who in 1950 first demonstrated that thermodynamic equilibrium between the neutron and proton populations is maintained through their collisions with electrons, positrons, and neutrinos. These 'collisions', of course, are not of the billiard-ball type, but are brought about by the weak interaction. Thus processes like the following, which go both ways, were constantly taking place as the universe cooled below $T_0 = 10^{13}$ K and the neutrons and protons became non-relativistic:

$$\nu_e + n \leftrightarrow e^- + p$$

$$\bar{\nu}_e + p \leftrightarrow \bar{\nu}_e + p$$

$$\nu_e + n \leftrightarrow \nu_e + n$$

Using the formula on page 81, we find that the ratio of the number densities of the neutrons and protons, N_n and N_p, at temperature T was

$$\frac{N_n}{N_p} = \exp\left(-\frac{1.5 \times 10^{10}}{T}\right).$$

Notice that the slight difference between the rest-mass energies of the neutron and the proton given in Table 3.2 is equivalent to an energy of 1.29 MeV, and hence to a temperature of 1.5×10^{10} K. At temperatures higher than this, N_n and N_p were almost equal. The ratio N_n/N_p dropped significantly, however, as the universe cooled below this temperature, until the actual numbers N_n and N_p themselves became so small that collisions became too infrequent to maintain thermodynamic equilibrium. Thereafter, the ratio dropped further for another reason: the one-way process of beta decay:

$$n \rightarrow p + e^- + \bar{\nu}.$$

The lifetime of a neutron with regard to this process is about 1,013 seconds, which is longer than the age of the universe at the time this process was initiated, but still not too large to be ignored.

When T drops so low that neutrons and protons are no longer in thermodynamic equilibrium, the above formula cannot be relied on to give the ratio N_n/N_p. Detailed book-keeping calculations of N_n and N_p taking into account the rates of the various reactions which influenced them have to be done by computer. The interesting point is that a significant number of bound light nuclei begin to emerge when N_n/N_p reaches the value of $\sim 1/7$.

Although the deuteron was the first to form, it was not the stablest nucleus, and it subsequently grew into bigger units through reactions like the following:

$$d + d \leftrightarrow {}_2^3\text{He} + n$$

$${}_2^3\text{He} + n \leftrightarrow {}_1^3\text{H} + p$$

$${}_1^3\text{He} + d \leftrightarrow {}_2^4\text{He} + n$$

The process terminates to all intents and purposes when the stablest of all light nuclei, the ${}_2^4\text{He}$ nucleus is formed. This nucleus has a binding energy of 28.3 MeV, and all neutrons are taken up in forming it. Since the helium nucleus has two neutrons and two protons, it is easy to estimate the proportion by mass, of baryonic matter, Y, which went to form ${}_2^4\text{He}$. Given a neutron to proton ratio of $1/7$ (see above), $Y \cong 1/4$. Y is called the mass fraction of helium: about a quarter of the total mass in the universe was thus in the form of helium. (Mass fractions for other nuclei are similarly defined.)

Before we proceed further, let us summarize the important steps so far:

1. The main synthesis of nuclei occurred in the temperature range 10^9–5×10^8 K, during which phase the universe aged from ~ 180 to ~ 700 seconds.

2. After cooling below $T_0 = 10^{13}$ K, the absolute numbers of baryons (n and p) dropped sharply, so their energy densities did not affect the dynamics of the universe significantly. The universal expansion rate was determined by leptons and photons. Of these, electron–positron pairs ceased to exist after the temperature dropped below $\sim 10^9$ K. Thus, during nucleosynthesis the dynamics of the universe were controlled largely by photons and neutrinos.

3. The eventual primordial proportion, Y, of $^4_2$He was determined by the ratio N_n/N_p at the onset of nucleosynthesis. This ratio is more or less uniquely determined (but see Chapter 5), thus giving a precise prediction of Y.

Further, the primordial nucleosynthesis cannot be continued beyond $^4_2$He in any significant way. A few light nuclei like d, $^3_1$H (triton), $^3_2$He, ^{7}Li, and ^{11}Be are formed, but in much smaller fractions than the $Y = 1/4$ for the $^4_2$He nucleus. This is because the nuclei after helium—lithium (Li), beryllium (Be), and so on—are not stable and revert to helium soon after they are formed. The stabler nuclei like carbon, oxygen, and so forth which lie beyond this gap of unstable nuclei, cannot be reached by this process of addition of neutrons and protons. It is in principle possible to produce carbon from three helium nuclei, as first pointed out by Fred Hoyle in 1954. However, after the first 3 minutes or so, the universe was not hot enough to bring about carbon production this way. None the less, the process was shown by Hoyle to be possible inside stars and to hold the key to further nucleosynthesis of heavier nuclei in stars.

Although it was produced in very small quantities (mass fraction $\lesssim 10^{-5}$), the deuteron holds important cosmological information. Its production depends very sensitively on the number density of baryons in the universe. If there were far too many baryons around when nucleosynthesis was taking place, it would be easy for collisions of deuterons with neutrons and protons to take place, leading to the destruction of all the deuteron and its conversion into helium. Thus survival of deuterons is favoured in a low- rather than a high-density universe. If we assume that most matter in the universe today is in the form of baryons, then the density parameter, Ω_0, can be linked with the primordial production of deuterons. For example, taking Hubble's constant as 75 km s^{-1} Mpc^{-1}, we find that in a universe with $\Omega_0 \gtrsim 0.2$, practically no deuterons are produced.

Nuclear Relics of the Big Bang

Although Gamow's original aim of producing all observed atomic nuclei in the early universe did not materialize, the work on primordial nucleosynthesis was revived in the 1960s. Researchers like Ya. B. Zeldovich, F. Hoyle, R. J. Tayler, P. J. E. Peebles, W. A. Fowler, R. V. Wagoner, and others repeated Gamow's calculations with increasing sophistication. The reason for this revival was that although stars were known to be likely sites for production of all nuclei from helium onwards, it was becoming increasingly apparent that stars cannot produce the light nuclei in the quantities observed. It is precisely in this area that the early universe appears to do the trick. Fig. 3.8 illustrates the estimated primordial abundances in Friedmann models according to the latest calculations.

Most striking in this context of course is the primordial production of helium, with a mass fraction Y of 0.25. A simple calculation based on the amount of starlight produced in the Galaxy from its birth to the present day provides an estimate of how much helium would have had to be synthesized in stars to generate that starlight; it comes out to a paltry value of $Y = 0.02$. This calculation of course uses the present luminosity of the Galaxy and an age estimate of ten billion years. If the

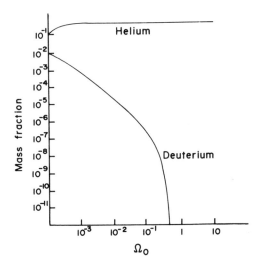

Fig. 3.8. The primordial abundances of helium and deuterium are plotted against Ω_0, the baryonic density parameter. Notice the sharp drop in the deuterium value for $\Omega_0 \gtrsim 0.2$.

Galaxy were considerably older or significantly brighter in the past than it is now, this estimate of Y would go up.

Actual observations of helium in several parts of the Galaxy yield values of $Y \sim 0.24-0.28$. From these, the stellar contribution has to be subtracted. Thus a primordial contribution of $Y \sim 0.25$ seems to be about right. Notice that this value of Y, unlike the deuteron abundance, does not depend sensitively on the baryon number density.

Ultraviolet observations of absorption lines in the spectra of bright stars indicate that the deuteron mass fraction, denoted by $X(d)$, lies in the range $9 \times 10^{-6}-3.5 \times 10^{-5}$. The Copernicus satellite, launched in 1973 to commemorate the quincentenary of the birth of Copernicus, gave the first reliable estimates of the abundance of deuterium in the universe. Since it is hard to come up with a stellar fusion process capable of making even this quantity of deuterium, and since the deuteron, if created primordially, could have been destroyed subsequently in galactic processes, we need big bang models of low Ω_0 to produce at least the observed $X(d)$. For a Hubble constant in the range 50–100 km s^{-1} Mpc^{-1}, the value of Ω_0 cannot exceed 0.0375–0.15. Thus, if most matter in the universe is baryonic, the low value of Ω_0 (<1) leads to the conclusion that the universe is open.

But this important, tight conclusion of the early 1970s has now developed a loophole! Whether the universe is open or closed depends not only on the density of baryonic matter, but on the overall density of matter. Thus we could have a low baryonic component of matter as above, together with non-baryonic matter (massive neutrinos, say) resulting in $\Omega_0 \geq 1$. In this case the universe would still be closed. Whether this loophole is really operative, we will discuss later.

Relic Radiation

It is evident from the preceding discussion that light nuclei now found in the universe can be looked upon as relics of the early hot phase of the universe in much the same way that the archaeological findings at Pompeii are relics of a once flourishing city which was destroyed by the volcanic activity of nearby Mount Vesuvius. Similarly, the currently observed microwave background is a relic of a once radiation-dominated universe.

We recall that after the electrons and positrons had annihilated one another substantially and after the universe had cooled below a billion degrees, say, only radiation and neutrinos remained to control the dynamics of the universe, with the former taking on the lion's share of the job. Although the matter component of the universe in the form of baryons and left-over electrons had no effect on the cosmic expansion, they did continue to interact with radiation, however, especially the electrons. (It is assumed here that there was an excess of electrons over positrons prior to annihilation. This assumption relates to a general excess of matter over anti-matter, and will be discussed in the next chapter.)

A typical electron scatters radiation very effectively. At low energies the scattering is by the so-called Thomson process, in which the energy of the incoming photon is not changed, although its direction of motion is. At high energies the scattering process is that known as 'Compton scattering', which results in the electron receiving a significant part of the photon's initial energy. Since the rest energy of the electron is ~0.5 MeV, the Compton process is not effective for photon energies much below this value, and hence would not have been significant in the universe after it had cooled below a billion degrees.

The Thomson scattering would have been quite effective, however, so photons would have been scattered frequently. If such a situation existed in the present state of the universe, it would be impossible to do astronomy! For photons from any astronomical source would not reach the telescopes of a remote observer intact: they would have been deflected too many times during their journey. That photons are able to make a journey of several billion light-years, bringing remote corners of the universe within the purview of our astronomical telescopes, is often expressed by the statement that the universe is 'optically thin'. The early universe with its frequent scatterings of photons by electrons, by contrast, was 'optically thick'.

Obviously, as the universe expanded and cooled, it changed at some stage from being optically thick to optically thin. When and how did this happen?

The answer to this question is provided, not surprisingly, by a process which removes free electrons from the cosmic brew. Remove the scatterers, and the radiation travels freely. The process occurred when the universe cooled to the temperature range of 3,000–4,000 K. At this temperature, free electrons combined with free protons to form electrically neutral hydrogen atoms.

This process is analogous to the binding of neutrons and protons into deuterium nuclei which took place earlier, at the much higher temperatures of 10^8–10^9 K. In that case the *nuclear* binding force arising from the strong interaction was able to trap neutrons and protons. The *chemical* binding

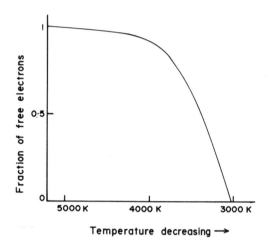

Fig. 3.9. A schematic plot showing how the number fraction of free electrons in the universe drops sharply as the universe cools from 5,000 K to 3,000 K, for Hubble constant of 75 km s^{-1} Mpc^{-1} and $\Omega_0 = 0.2$.

between an electron and a proton is electrostatic in nature, and therefore much weaker; compared to the binding energy of the deuteron (2.22 MeV), the binding energy of the hydrogen atom is very small (only 13.59 eV). Thus, for electrons and protons to be trapped by this force, their speeds must be considerably smaller, and hence their temperature considerably lower, than in the earlier phase of the formation of the deuteron. Calculations show that, depending on the number density of electrons (which in turn can be related to the number density of protons *now present* in the universe), the bulk of the electrons were trapped into forming hydrogen atoms during the cooling of the universe from 4,000 to 3,000 K. Because this process is known as 'radiative recombination', this epoch is often called the 'recombination epoch'. Fig. 3.9 shows how the fraction of free electrons drops as the universe cools through this crucial phase.

In other words, by the time the universe had cooled to 3,000 K, it had become optically thin. Radiation then became decoupled

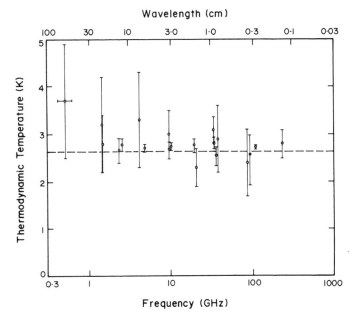

Fig. 3.10. A composite spectrum of the microwave background combining the data at various frequencies. The dashed curve corresponding to the black-body spectrum of T = 2.7 K is drawn through the data points and their error bars.

from matter in much the same way that the neutrinos had earlier become decoupled from the other constituents of the universe. And like the neutrinos, the decoupled radiation preserved its equilibrium distribution of black-body radiation even afterwards, although with a temperature which fell as $1/S$ as the universe expanded. It is therefore to be expected that the relic radiation observed today should have a black-body distribution.

As mentioned earlier, this expectation appears to be borne out by recent observations of the microwave background at different frequencies. Fig. 3.10 presents the latest data on this count. The effective background temperature of about 3 K today tells us that the decoupling of radiation from matter took place at the epoch in which the red-shift was ~1,000.

The reader will recall that this particular epoch featured earlier, but in a different context! For it was at this epoch that the universe switched over from being radiation-dominated to becoming matter-dominated. *A priori* there is no reason why both events should have occurred at approximately the same epoch. Nor does this coincidence find a ready explanation in the big bang scenario developed so far.

Perhaps we could understand the significance of the coincidence better if we knew why the present-day temperature of the microwave background is 3 K. Why is it not 7 K, as supposed by Gamow in 1953, or 5 K as proposed by Alpher and Herman in 1949? Although early universe calculations predict an almost unique value for the helium mass fraction, Y, they are not able to predict a value for the present-day temperature of the relic radiation.

This shortcoming is often expressed by observing that the photon-to-baryon ratio in the universe is not determined by the above calculation. At a temperature T_0, the formulae on page 78 tell us that the number of photons per unit volume should be

$$N_\gamma = 19.2\pi\left(\frac{kT_0}{ch}\right)^3.$$

Thus, for $T_0 = 3$ K, the photon number density is 545 per cm^3. If we assume that of all the matter present in the universe a fraction, f_B, by mass is in the form of baryons, each with a typical mass m_p (the mass of the proton) then from the formulae on page 62, we see that the number density of baryons in the universe at present is

$$N_B = \frac{3H_0^2\Omega_0}{8\pi G m_p}f_B.$$

After substituting numerical values for the various physical constants, we can find the ratio of N_γ to N_B. Since the value of Hubble's constant is not known exactly, it is customary to express it as $H_0 = 100\,h_0$ km s^{-1} Mpc^{-1}, where the value of the dimensionless number, λ_0, is believed to lie between 0.5 and 1. We then get

$$\frac{N_\gamma}{N_B} = 4.57 \times 10^7 \, (\Omega_0 f_B h_0^2)^{-1} \left(\frac{T_0}{3}\right)^3.$$

For example, for $\Omega_0 = 1$, $f_B = 0.1$, $h_0 = 0.75$, $T_0 = 3$ K, this ratio is about 8×10^8. It has remained unchanged from the early epochs, since both the photon number density and the baryon number density scale as S^{-3}. Obviously the key to this ratio must lie in the history of the universe prior to the epochs we have discussed in this chapter. What was the universe like at those earlier epochs?

Until the mid-1970s, this question was considered unanswerable. For, to describe the physical behaviour of the universe, we must have adequate physical laws. The laws of physics as understood in the mid-1970s did not fill the bill. The situation has changed dramatically since then, however, with the arrival on the scene of various grand unified theories. These theories offer the hope not only of understanding the various initial conditions assumed in the work described in this chapter, but also of giving a physical description of the universe at epochs considerably closer to the singular epoch of the big bang.

It is to GUTs, therefore, that we must next turn our attention.

4

Towards the Unification of Physics

> There is no excellent beauty that hath not some
> strangeness in the proposition.
>
> Francis Bacon

The last chapter brought us to a tantalizing situation in our probe of the early history of the universe. The microwave background and the evidence of the abundance of light nuclei in the universe seem to support the picture of the hot big bang first mooted by George Gamow. However, the type of calculations we described were relevant to the early universe only when it was between 1 second and 3 minutes old. The success of those calculations naturally prompts present-day theorists to be yet more daring, and to approach the singular big bang epoch even more closely.

It is not solely the adventurous instinct which guides the cosmologist in this quest. There are questions to be answered about the universe which are of such a fundamental nature that their answers most probably lie in the very early phase when the universe was even younger than 1 second. We raised some of these questions in Chapter 3. When and how was the material composition of the universe decided? Is it largely made of matter, rather than antimatter? And if so, why? What determined the relative proportion of radiation to matter (the ratio N_γ/N_B)? How did the nucleons come into existence? ... Questions like these remain to be answered. Many cosmologists also believe that the seeds of the large-scale structure found in the universe, the galaxies, clusters, and superclusters, are to be found in primordial epochs.

However, new concepts are needed if the theorist is to answer these questions, and for a while during the 1970s, it looked as if basic physics could not deliver them. But this situation changed dramatically towards the end of the decade. The 1970s saw the electroweak theory of Steven Weinberg, Abdus Salam, and Sheldon Glashow gain acceptance. Encouraged by the success of this theory, theoretical physicists began to look for an even wider framework that would also incorporate the strong interaction in a unified theory, their ultimate hope being to unify *all* known basic interactions of physics. The so-called grand unified theories (GUTs) and the even more esoteric supersymmetry (SUSY) and supergravity theories are examples of such attempts at unification.

In this chapter we will familiarize ourselves with the essential features of some of these ideas, for therein lie the hopes of the cosmologist wishing to probe the very early universe. As we will discover, the benefit does not flow one way only: the theoretical physicist wishing to try out his pet unification theory badly needs the cosmologist!

On Particles and Fields

The concept of 'interaction' in physics is inextricably mixed with the concept of the structure of matter; for the former describes how different components of the latter behave towards each other. History shows that our understanding of the former often goes hand in hand with our understanding of the latter.

The electromagnetic theory, for example, provided the explanation of the chemical properties of matter. The periodic table of chemical elements received theoretical justification when the atomic structure of these elements was studied with the help of the electromagnetic theory in a quantum framework. Conversely, the origin of electric and magnetic fields, and their apparently strange manifestations like diamagnetism, paramagnetism, and so on could be understood only by studying the microscopic structure of matter.

Here we will not take the historical route to our present understanding of matter, interesting though it is to see how haphazardly our knowledge has been built up. Rather, we will summarize the highlights of what is known (or believed to be known!), our description in this section being empirical. In later sections we will see how theorists see a pattern amidst the bewildering range of properties of subatomic particles, and how they seek to build on it further in their attempts to achieve a single unified picture.

According to the classical picture of the electromagnetic interaction, two charged particles interact with the intervention of an electromagnetic *field*. Take two electrons, for example. When one of them moves, it disturbs the ambient field. The ripples created in the field travel outwards at the speed of light until they 'hit' the other electron. The interaction between the two electrons is complete when energy and momentum have been exchanged between them, the 'carrier' of these being the field. Thus the electromagnetic field itself has energy and momentum, which are transmitted from one point to another by waves travelling at the speed of light. This field–particle interaction picture is described by the equations of Maxwell and Lorentz.

Quantum theory altered this picture significantly, in two ways. First, the distinction between field and particle disappeared. The classical electromagnetic field was visualized as a continuum occupying space; in the quantum version it acquired graininess. The 'grains' of the field are the photons, the particles of light. The energy and momentum residing in the electromagnetic field in a given volume can be identified with those of an assembly of photons in that volume. When two electrons interact, they do so by exchanging a photon, which transmits energy and momentum from one to the other. Each photon is endowed with a specific 'quantum' of energy and momentum.

The electrons likewise can be visualized as quanta of a field. In quantum theory of the electromagnetic field and the electron field one talks of states of the latter containing zero, one, two, or however many particles, and of how these states change because of the interaction of the electron field with the electromagnetic field. In other words, one does not talk of individually identifiable electrons interacting with one another as in the classical picture of Maxwell and Lorentz, but of many particle states of electrons and photons interacting with one another.

This change of description gave rise to the second important difference from classical theory. A 'vacuum' in classical physics is devoid of *anything*. In quantum physics, application of the uncertainty principle to fields and their particle quanta changed the notion of vacuum. According to quantum theory, 'vacuum' is the state of lowest energy, but it is not devoid of anything. For example, a classical electron vacuum means simply that there are no electrons. A quantum vacuum of electrons, on the other hand, is full of virtual pairs of electrons and positrons. A virtual pair in general implies a particle–anti-particle pair which is spontaneously created and destroyed. So although the net effect is zero, such pairs, in spite of their short lives ($\sim 10^{-23}$ s), can produce tangible effects on real interacting particles.

Thus the field–particle duality and the non-trivial status of the vacuum which follows from quantum field theory make *quantum electrodynamics* (abbreviated QED) a considerably richer theory than classical Maxwell–Lorentz electromagnetic theory. The success of QED in explaining diverse experimental phenomena in the laboratory has prompted physicists to adopt the quantum field-theoretic description for other interactions too. (There is a further feature of QED—namely, that it is a *gauge theory*—which also appears to be a universal aspect of physical interactions; but more of that later.)

Since field theory talks of particles as collections of quanta of energy and momentum, there is an underlying 'sameness' about particles of the same species. The particles are not individually identifiable, and a given state of the field describing a collection of particles is not changed if particles are interchanged arbitrarily. In mathematical language, this statement translates into the following: the wave function describing the quantum state of the group of particles is either symmetric or antisymmetric with respect to the interchange of any pair of particles. For some particle species like electrons and protons the 'antisymmetric rule' applies; for example, if the particles are labelled 1, 2, 3, ..., then their wave function,

ψ, satisfies the condition $\psi(1, 2, \ldots P, \ldots R, \ldots) = -\psi(1, 2, \ldots R, \ldots P, \ldots)$, whenever any pair of particles, P, R, are interchanged. In quantum theory ψ is a complex function, and $|\psi|^2$ denotes the probability density for the quantum state. For other particle species like photons the 'symmetric rule' holds, and for the interchange of P and R, $\psi(1, 2, \ldots P, \ldots R, \ldots) = +\psi(1, 2, \ldots R, \ldots P, \ldots)$.

Notice that if the dynamical details specifying the particle P—for example, the energy, momentum, and so on—are the same as those specifying R, then in the antisymmetric case ψ becomes zero. This means that for particles obeying such a rule, no two can be in the same state. This is known as the 'Pauli exclusion principle'. Wolfgang Pauli, who first enunciated it in 1925, also found a deeper connection between the above symmetry/antisymmetry rules and a dynamical property of the corresponding particles.

This property is known as the 'intrinsic spin' of the particle. The notion of spin has a classical origin, of course, typical examples being spinning tops, carousels, the earth and other planets spinning about their polar axes, and so on. Intrinsic spin is a quantum-mechanical property, however, and it is measured by the angular momentum of the subatomic particle in units of $\hbar = h/2\pi$. Thus, when we say that an electron has spin 1/2 or that the photon has spin 1, we mean that the angular momenta of these particles are $\hbar/2$ and $\hbar$ respectively. Notice, however, that intrinsic spin can take only certain discrete values. Thus, if an experiment is set up to measure the spin of an electron in *any* given direction, the answer will always be either $\hbar/2$ or $-\hbar/2$. The process of measurement allows the electron to exist only in one or other of these two possible spin states. This is the property which sets the intrinsic spin apart from the classical notion of spin.

The magnitude of the spin of the particle has implications for the directionality of the corresponding field. For example, spin 1 corresponds to a *vector field*: the photon has spin 1 because the electromagnetic field potential transforms as a vector—that is, as a quantity with both magnitude and direction. Likewise spin-2 particles are associated with fields which behave like tensors of rank 2. (An example of a tensor of rank 2 is the stress that develops in a body when it is twisted. Two directions are needed to specify a tensor—in the above example, the direction of the twist and the direction of the stress force.) The vectors and tensors may be looked upon as ways of expressing the invariance of physical laws at a given point under the Lorentz transformations (see Chapter 2). The mathematical representation of invariance can also be expressed by so-called spinors. Spinor fields describe particles of half-odd integer spin (1/2, 3/2, and so on). We will return to this correspondence later.

The deeper connection discovered by Pauli is between spin and the property of symmetry/antisymmetry. For particles with integral spin—for example, mesons with spin 0, photons with spin 1, gravitons with spin 2, and so on—the symmetric property holds. As we saw earlier, such particles are called 'bosons'. For particles with spin measurable in half-odd integral values—for example, electrons with spin 1/2, Δ-particles with spin 3/2—the antisymmetric rule applies, and such particles are called 'fermions'.

These names are given on the basis of the rules of statistics which these particles follow. Suppose we have N particles of the same species. How can they be distributed among M available dynamical states? If $N > M$, then at least two particles will have the same dynamical state, which is forbidden for fermions by the Pauli exclusion principle. Thus no solution is possible for fermions, although it is for bosons. Even when $N < M$, the combinations allowed by the Pauli principle for fermions are not the same as those for bosons. And both fermions and bosons differ in this respect from classical macroscopic particles.

The rules of statistics for fermions were first investigated by Enrico Fermi and Paul Dirac, while those for bosons were first highlighted by Satyen Bose and Albert Einstein. The validity of the Fermi–Dirac and the Bose–Einstein statistics has been amply demonstrated in physics in such diverse phenomena as

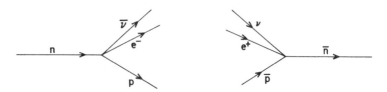

Fig. 4.1. On the left we have a neutron decaying into a proton, an antineutrino, and an electron. Under CPT we get, as shown on the right, an antiproton combining with a neutrino and a positron to produce an antineutron.

black-body radiation, superconductivity, the distribution of electrons in white dwarf stars, and that of neutrons in neutron stars.

Finally, the notion of matter and antimatter can itself be traced to the way the two types of quantum statistics operate. Again, we will not go into mathematical details, but will summarize the result, which is clear and simple: all fermions of matter have antimatter counterparts, whereas bosons are symmetric with respect to matter and antimatter. Thus the antimatter counterparts of the electron (e^-) and the proton (p) are respectively the positron (e^+) and the antiproton ($\bar{p}$), whereas the photon treats the two species symmetrically. Matter and antimatter brought together annihilate each other, producing radiation; hence $e^- + e^+ \rightarrow \gamma + \gamma$. Given sufficient intensity of photons, as, for example, in the early universe (see Chapter 3), the reverse reaction can also take place.

The basic symmetry between the behaviour of matter and antimatter as given by quantum field theory is expressed by the so-called CPT theorem. This theorem, illustrated in Fig. 4.1, states that, corresponding to any reaction involving sub-atomic particles, we can obtain another in which the particles are replaced by antiparticles (C, equals charge conjugation); the spatial parity (P) is reversed—that is, we replace particles by their reflected mirror images—and the direction of time (T) is reversed so that what went *in* the reaction becomes what came *out*, and vice versa.

A crude analogy of CPT transformation is as follows.

Imagine a country which decides to switch its traffic flow of road vehicles from the left-hand side (as in the United Kingdom, for example) to the right-hand side (as in the United States). The motorways will correspondingly change their entry points to exit points, and vice versa.

Particle physicists have encountered situations in which different components of CPT are violated, such as parity or time reversal, but the combination CPT itself is not violated. In other words, nature has so far not revealed any interaction whose CPT counterpart does not also exist.

The Electroweak Interaction

The unification of the electromagnetic interaction with the weak interaction provides the stepping-stone to all further attempts at unification. We will first try to understand pheno-menologically what is involved in arriving at the electroweak interaction. Again, the example of the electromagnetic inter-action will help.

The initial experiments of Coulomb led to the postulation of two separate interactions, one of electrical attraction/repulsion between unlike/like electric charges, the other of a force be-tween two magnets. Given only these experiments, it is not possible to conjecture that the two phenomena might be differ-ent aspects of a unified electromagnetic interaction. But later experiments of Ampère and Faraday and others did give hints of a unified interaction, in that electric charges flowing through a wire were found to deflect a magnetic compass needle, and a flow of electric charges could be generated in a wire by moving a magnet rapidly through it. Experiments like these finally led to the unified electromagnetic theory of Maxwell.

Now let us consider two typical examples of the weak interaction:

$$n \rightarrow p + e^- + \bar{\nu}$$

and

$$e^- + \nu \rightarrow e^- + \nu.$$

The first describes beta decay of a neutron, which we encountered in Chapter 3, whereas the second describes electron–neutrino scattering. Prima facie, there is no hint here of any connection with the electromagnetic interaction, since the strengths of the interactions do not depend on electric charge.

Early formulations of the theory of such weak processes during the 1950s, however, required electric charge to be exchanged between the interacting particles. Thus, in electron–neutrino scattering, the theory required the electron to become a neutrino and the neutrino to become an electron. Likewise, in the beta decay, the neutron becomes a proton. If the weak interaction were completely separate from the electromagnetic interaction, how could these processes apparently 'take note of' electric charge?

To answer this question and to follow the hint contained in it for unifying the two interactions, let us first consider leptons only. As we saw in Chapter 2, leptons are particles immune from the strong interaction. So far three pairs of leptons are known: e^-, v_e; μ, v_μ; and τ, v_τ. The way we have written these pairs, the first members (the electron, the muon, and the τ-lepton) have the same electric charge, whereas the second, the corresponding neutrino, is electrically neutral. But there is more to it than just this distinction. The reactions (weak and electromagnetic) conserve the so-called lepton number, L, for *each pair*. In counting the lepton number, the antiparticle is given the value $L = -1$. Thus in the muon decay $\mu^- \to e^- + \bar{v}_e + v_\mu$, the left-hand side has muon lepton number $L = 1$, as does the right-hand side. The electron and τ-lepton numbers are zero. Thus, a decay such as $\mu^- \to e^- + \bar{v}_e + v_e$, in which the lepton number is *not* conserved for each pair, is forbidden.

The three pairs are called the three 'flavours' of leptons. It seems that the two members of a pair always go together as far as the weak interactions are concerned. The neutrinos, being chargeless, do not participate in the electromagnetic interaction, of course. There is another difference between the electromagnetic and the weak interactions: the former pre-

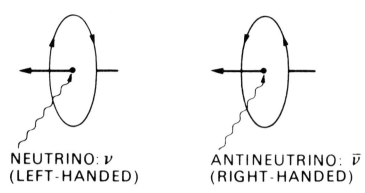

NEUTRINO: ν
(LEFT-HANDED)

ANTINEUTRINO: $\bar{\nu}$
(RIGHT-HANDED)

Fig. 4.2. If parity were conserved, then the mirror image of the left-handed neutrino should have been found. But it has not been. The right-handed particle shown on the right is the antineutrino.

serves parity, whereas the latter does not. Before proceeding further, let us familiarize ourselves with parity violation.

As mentioned earlier, if parity is conserved in a given interaction, then for each reaction seen in nature, the mirror reflection should also be possible. Violation of parity can be seen from the way in which particles spin with respect to the direction of their motion. Take the electron, for example. It has two spin states, with spin values $\pm 1/2$ when measured along the direction of motion. Now we say that the electron is spinning in a left-handed fashion when its spin is $-1/2$ and in a right-handed fashion when its spin is $+1/2$, as measured along the direction of motion. (If we view the spin along the direction of motion, clockwise spin is positive and anticlockwise spin is negative.) A neutrino, however, has only one state, that of left-handed spin. Moreover, in weak interactions only left-handed electrons and neutrinos are found to participate.

The statement that 'God is left-handed' owes its origin to this circumstance. The mirror reflections of such interactions, which involve right-handed neutrinos, do not exist in nature (see Fig. 4.2). Such parity-violating electrons can be used to inform a distant extraterrestrial intelligent being about what we mean by right and left: we simply tell him to perform an experiment involving weak interactions and to measure the sign of the spin of the emerging neutrinos or electrons!

Electrons participating in the electromagnetic interaction respect parity conservation, however, and are found in two states, right- and left-handed. We can therefore 'subdivide' the electron–neutrino combination into two groups of states:

$$\begin{bmatrix} e^- \\ \nu_e \end{bmatrix}_L \text{ and } (e^-)_R$$

where L and R denote left- and right-handed states. Notice that the neutrino has *no* right-handed state. Only the left-handed doublet combination participates in the weak interaction. We will consider it first.

It is convenient for this purpose to describe the particles as quantum-mechanical states with appropriate wave functions. Thus, instead of saying (in the language of classical physics) that the electron is at such and such a location, we talk of a wave function for the electron, ψ_e, which depends on space and time. The value of $|\psi_e|^2$ at any point P tells us the probability density—that is, the probability of finding the electron in a unit volume of space in the neighbourhood of P (note that, in general, ψ_e is a complex number in the mathematical sense). For the electron–neutrino pair we write the wave function, ψ, as a pair:

$$\psi = \begin{bmatrix} \psi_e \\ \psi_\nu \end{bmatrix}_L.$$

Thus ψ has two components, ψ_e for the electron and ψ_ν for the neutrino, both of which are left-handed.

Now, here as elsewhere in quantum mechanics, the *principle of superposition of states* is expected to operate. So in seeking a unified theory, we will adhere to this principle, which in the present context means the following. Since the weak interaction treats the electron and the neutrino alike, there is no sharp distinction between ψ_e and ψ_ν. In other words, if we took linear combinations of these like

$$\psi'_e = U_{11}\psi_e + U_{12}\psi_\nu \text{ and } \psi'_\nu = U_{21}\psi_e + U_{22}\psi_\nu,$$

then the new combination,

$$\psi' = \begin{bmatrix} \psi'_e \\ \psi'_\nu \end{bmatrix}_L,$$

should also represent possible states of the electron and the neutrino.

The coefficients U_{11}, U_{12}, and so on are constants which may take complex values, since the ψs are also complex functions. It is convenient to group them into a matrix:

$$U = \begin{Vmatrix} U_{11} & U_{12} \\ U_{21} & U_{22} \end{Vmatrix},$$

and to write the new wave function, ψ', as the product $U\psi$.

At this stage we use the physical interpretation of ψ— namely that $|\psi|^2$ denotes the probability density. In other words, $|\psi|^2 = |\psi_e|^2 + |\psi_\nu|^2$ denotes the probability density of finding the combination in the state described by ψ. If we demand that this remain unchanged when we go over to the new way of looking at the same combination, then $|\psi|^2 = |\psi'|^2$. This requirement translates into the condition for U that $UU^* = U^*U = 1$, where U^* is the so-called Hermitian complex of U:

$$U^* = \begin{Vmatrix} \overline{U}_{11} & \overline{U}_{21} \\ \overline{U}_{12} & \overline{U}_{22} \end{Vmatrix}.$$

What we have done here is analogous to rotation in a plane, as shown in Fig. 4.3. If we choose Cartesian axes to describe the components of a vector ℓ as X and Y respectively, then the magnitudes of X and Y will change as we rotate our coordinate system in the plane. But the vector itself is the same; in particular its magnitude, ℓ, is unchanged. Thus, $\ell^2 = X^2 + Y^2$ is the same for all X and Y. We may likewise look upon the above transformation as some kind of rotation in an abstract space (sometimes called the 'weak isospin space').

Those familiar with matrix algebra will recognize the matrix U as what is called a 'unitary matrix'. From its product rule

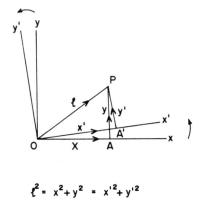

$$\ell^2 = x^2 + y^2 = x'^2 + y'^2$$

Fig. 4.3. Under rotation of coordinate axes, the components, X, Y, of a vector change, but its magnitude, ℓ, is not changed. Likewise the two component wave functions describing a state may change under spinor rotation, but the total probability of the state remains unchanged.

given above, it is easy to deduce that it has a determinant whose modulus is one. However, we can make a more definite statement about U by noting that the physical content of ψ is not altered if we multiply it by a complex number of unit modulus. Such a number is best written as $\exp(i\theta)$ where θ is a real number. Thus by writing $\psi\exp(i\theta)$ instead of ψ for constant θ, we are not changing any physical description of the system. (For example, the probability density remains unchanged.)

This transformation is called a 'phase transformation', θ being the phase. We will encounter it again in later parts of this chapter in a more significant way. For the time being we note the latitude it gives us in specifying ψ and ψ'. This latitude can be used in adjusting ψ, and hence the phase factor in U, to ensure that the determinant of $U = 1$. In other words, physically significant transformations of ψ can be described by those unitary matrices whose determinant is unity. Such matrices are called 'special unitary matrices'.

These transformations form a group called the 'special

unitary group of order 2' (since it refers to 2 × 2 matrices), which is written as SU(2). Although we have arrived at this group by considering the electron–neutrino pair, historically speaking it was first introduced into particle physics by Heisenberg, in 1932, in the context of the proton–neutron pair. His reasons were similar to ours—namely, that the strong interaction treated the pair symmetrically, even though the proton has electric charge and the neutron is electrically neutral. In the context of the proton–neutron pair this symmetry is called the 'isospin symmetry'.

The concept of invariance with respect to phase transformation which we have discussed for the weak interaction is also relevant in a simpler form to the electromagnetic interaction. In this case we have only the electron wave function to deal with, since the neutrino has no charge. The transformations are therefore described by 1 × 1 matrices—that is, by pure numbers; thus $\psi_e \rightarrow U\psi_e$, where $U = \exp(i\theta)$. The numbers U form a unitary group of order 1, written U(1). (Notice that the requirement of phase invariance here does not lead us to SU(1), since that group is the trivial one containing only one element, $U = 1$. Indeed, the fact that we were able to adjust the phases to ensure that the determinant of $U = 1$ in the case of the SU(2) group was because we had two adjustable phases of ψ_e and ψ_ν at our disposal. Here we do not have that freedom, and so we cannot arrange to have $U = 1$.)

The U(1) group is abelian, since its elements $\exp(i\theta)$ commute with one another. This is not the case with the elements of the SU(2) group, which is non-abelian.

There is a particularly nifty way of describing special unitary matrices of any order. We will illustrate it for order 2. Just as a complex number of unit modulus is expressible as $\exp(i\theta)$ for real θ, so we may write a 2 × 2 unitary matrix as $U = \exp(iH)$, where H is a Hermitian matrix. That is, the components of H satisfy the conditions H_{11}, H_{22} real; $\overline{H}_{12} = H_{21}$.

In the case of a complex number $U = \exp(i\theta)$, the phase $\theta = 0$ corresponds to U having the value unity. Likewise, the matrix U has unit determinant if the matrix H has zero trace— that is, if $H_{11} + H_{22} = 0$. It is easy to see therefore that only three real numbers, θ_1, θ_2, and θ_3, are needed to specify H, and hence U:

$$H_{12} = \theta_1 + i\theta_2; \quad H_{21} = \theta_1 - i\theta_2; \quad \text{and} \quad H_{11} = -H_{22} = \theta_3.$$

It is convenient to write H in the form

$$H = \theta_1 \sigma_1 + \theta_2 \sigma_2 + \theta_3 \sigma_3,$$

where σ_1, σ_2, and σ_3 are 2×2 matrices:

$$\sigma_1 = \begin{Vmatrix} 0 & 1 \\ 1 & 0 \end{Vmatrix}, \quad \sigma_2 = \begin{Vmatrix} 0 & -i \\ i & 0 \end{Vmatrix}, \quad \sigma_3 = \begin{Vmatrix} 1 & 0 \\ 0 & -1 \end{Vmatrix}.$$

These matrices are often called the 'Pauli matrices', since they were first introduced by Wolfgang Pauli to describe the intrinsic spin of the electron.

Where has all this algebra taken us? We started by considering the transformations of the electron–neutrino wave function, bearing in mind the symmetry of the two particles with respect to the weak interaction. From the symmetry arguments we arrived at the SU(2) group, which can be described essentially by three real parameters θ_1, θ_2, and θ_3. Into this mathematical picture we now introduce some physics.

Let us ask ourselves how an electron state can change to a neutrino state, and vice versa. Since an electron has a negative charge and a neutrino has zero charge, a charged particle will have to be exchanged for this to happen. Going purely by analogy with electrodynamics, in which a photon in exchanged between charged particles, we assume the exchanged particles to be bosons, denoted by $W^{\pm}$. Thus we may write

$$e^- \rightarrow W^- + \nu_e \quad \text{and} \quad \nu_e \rightarrow W^+ + e^-.$$

Like photons, W-bosons have spin 1; but unlike photons, they are massive and carry electric charge, since an exchange of charge was required by the weak interaction theory of the 1950s.

When we compare the change of states brought about by charge exchange with the SU(2) transformations, we see a link between the W-bosons and the matrices σ_1 and σ_2. But what about σ_3? Since σ_3 is a diagonal matrix, it corresponds to no

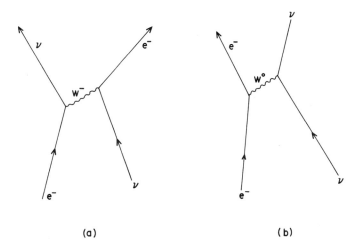

(a)   (b)

Fig. 4.4. The W-bosons mediate in weak interactions. In (a) the W^- boson is charged, and an electric charge $(-)$ is exchanged. In (b) no charge is exchanged, and the W-boson is neutral.

interchange of ψ_e and ψ_ν. Thus there is no charge exchange. In principle we could accommodate this fact by introducing a neutral particle W^0 such that

$$e^- \to W^0 + e^- \quad \text{and} \quad \nu_e \to W^0 + e^-.$$

Such a neutral W-boson was not included in the framework put forward in the 1950s, however. In fact, a new theory was needed to accommodate all three W-bosons dynamically as per the SU(2) model. The role of these bosons is illustrated in Fig. 4.4.

Before thinking of a dynamical theory, let us first try to bring the electromagnetic theory into this picture. Recall that the SU(2) group above acts on the left-handed wave functions $(\psi_e)_L$ and $(\psi_\nu)_L$ only. Recall also that both $(\psi_e)_L$ and $(\psi_e)_R$ have electromagnetic interactions which obey the U(1) group. Crudely speaking, therefore, we would expect the two interactions together to be described by a 'product' of the two groups—namely, SU(2)$_L$ × U(1). Here the subscript L on SU(2) indicates that it acts on left-handed components only.

This reasoning can be refined by adjusting U(1) for the fact that the left- and right-handed parts of the electron wave function do not behave in the same way under SU(2)$_L$; but we will not go into those details here.

The effect of such a combined group is to introduce *two* neutral bosons into the picture, the photon as well as the neutral boson W^0. The same four bosons, W$^+$, W$^-$, W^0, and γ are expected to mediate the electroweak interactions of *all* leptons, regardless of flavour. Thus the SU(2)$_L$ × U(1) group would act on the pairs (μ^-, ν_μ^-) and (τ^-, ν_τ) also.

The picture developed so far does not give us a theory, but only the hint of a theory. Indeed, a theory that unifies the electromagnetic and the weak interactions must answer several questions which emerge from the above discussion.

First, the range of the electromagnetic interaction—that is, the distance over which its effect is felt—is infinite, while that of the weak interaction is finite and very small—~10^{-12} cm. Theoretical considerations lead us to the conclusion that the range, r, of an interaction varies inversely as the mass, m, of the mediating particle:

$$r \sim \frac{\hbar}{mc}.$$

The long range of the electromagnetic interaction is accounted for by the photon having zero mass. The short range of the weak interaction is similarly due to its mediating particles, the W-bosons, having large mass (how large we shall see later!). How and when does such mass distinction arise in a unified picture?

The two interactions are not of comparable strength at the moderate energies obtainable in a typical laboratory. As its name implies, the weak interaction is considerably weaker than the electromagnetic one. Again, we will not go into details of how this relative weakness is measured; we simply state that at first sight the difference between the strengths of the two interactions appears too large to be accommodated naturally in a unified picture.

The third point is more technical, and relates to the property

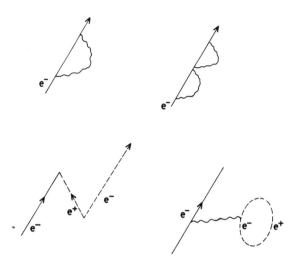

Fig. 4.5. In QED even a 'free' electron is not a simple concept. Shown here are a few of the various ways an electron (straight line) can interact with photons emitted and absorbed (curly lines) or with pairs of electrons and positrons created and annihilated in a vacuum (shown by dashed lines). The mathematical calculations regarding these processes lead to infinite results, which can be eliminated by techniques of renormalization.

of a field theory called 'renormalizability'. In QED, as we noted earlier, even the notion of a single charged particle like an electron moving in a vacuum is not simple. Fig. 4.5 indicates the complexity of what goes on. The electron keeps interacting with particle–antiparticle pairs which are being spontaneously created and destroyed in a vacuum. If all these processes were properly taken into account, the mathematical calculations would lead to infinite values for such physically measurable quantities as probability, mass, electric charge, and so on. These infinities can, however, be handled in a meaningful way to yield finite answers. A mathematician demanding rigour will baulk at taking such liberties, but a physicist looking for measurable numbers can be satisfied with such a technique, provided the rules for handling infinities

fulfil a set of clearly laid-out objectives. Two main require-
ments are that the rules of subtracting infinities from one
another be unambiguously specified, and that they lead to
answers in very good agreement with experiments.

This technique of removing infinities is called 'renormalization'. Although it works for QED, it does not work for
every field theory. The reason why QED is renormalizable is
connected with the photon having zero mass. We have already
mentioned that the photon is a boson of spin 1. A *massive*
particle of spin 1 can exist in three independent states of spin,
one parallel to its direction of motion, one opposite to that
direction, and the third perpendicular to it. It is this last state
which is not found in particles of zero mass such as the photon.
When the mathematical expressions of various field-theoretic
processes are written out, it is found that the infinities arising
.from the third (perpendicular) spin state cannot be removed
by renormalization.

Now we have already seen that the W-bosons mediating the
weak interaction are massive. Moreover, they have spin 1. It
therefore follows that a field theory for the weak interaction
developed along the lines of QED will be non-renormalizable.
How can we combine such a theory with renormalizable
QED?

Gauge Theories and Spontaneous Symmetry Breaking

These difficulties, insurmountable though they seem, receive
a solution through a technique which has now gained considerable acceptance among theoretical physicists, largely
because it led to the successful construction of an electroweak
theory. To understand this technique, we first recall Chapter 1
and Aristotle's penchant for circles as natural trajectories of
motion.

Aristotle's choice of circles was based on their symmetrical
geometrical structure. But Aristotle also conceded that, as
opposed to natural motion along circles, there was violent
motion which departed therefrom. In other words, the

symmetry preferred by nature was broken when external (for example, man-made) agencies interfered through the application of forces.

Although it turned out that Aristotle was incorrect in his ideas about dynamics, his stress on symmetry finds an echo in modern theoretical physics. We have already encountered it in cosmology in the Robertson–Walker models. In special relativity the principle that all inertial observers find the same description of physical laws is an expression of symmetry. The Lorentz transformations connecting the space–time coordinates of two inertial observers form a group called the 'Lorentz group', under which the laws of physics are believed to be invariant. It was the symmetry in behaviour between the electron and the neutrino in the weak interaction which led us to the SU(2) group.

However, it sometimes happens that the symmetry in nature's behaviour is not an obvious one; indeed, it may be hidden under certain circumstances. It may happen, for example, that symmetry arguments allow several alternatives to a system, which may spontaneously select one of them. After selection it would therefore appear that the symmetry is not there are all.

The classic Greek story of Buridan's ass illustrates this idea. Two piles of grass were placed symmetrically on each side of the ass, one to the right, one to the left. Because of the perfect symmetry, there was no reason for the ass to select one pile in preference to the other, and as a result it starved to death. If the ass had learned modern particle physics, however, it would have survived! For, by spontaneously breaking the symmetry between right and left, it could have selected one of the piles and satisfied its hunger.

Notice that the ass's position in the centre is unstable. A slight perturbation (such as a shake of its head to avoid a fly) could have altered its state minutely to make one pile appear closer than the other, and the decision would have been made.

An equivalent example of spontaneous symmetry breaking is found in the ferromagnet, a magnetized bar of iron. The equations which govern the behaviour of iron nuclei and the electrons which go around them are invariant under rotation. This symmetry means that the free energy of the bar after it has been magnetized is the same whether one end is made the north pole or the other. At high temperature, the energy of magnetization is least when the alignment of the electrons relative to the nuclei is random. The energy curve is therefore symmetric with respect to the direction of magnetization (that is, it does not depend on which end is made the north pole). The bar 'prefers' the stable state of least energy, however, when it is not magnetized. But as the temperature is lowered, the energy of magnetization behaves strangely. It maintains the symmetry with respect to the direction of magnetization, but, below a certain temperature called the 'Curie temperature', which equals 770°C for iron, the manifestly symmetrical state of zero magnetization is not the one with least energy. (See Fig. 4.6.) The energy curve below this temperature shows two minimum states symmetrically situated with respect to the two ends of the bar, while the state of zero magnetization has a higher energy. The iron bar, like Buridan's ass, is therefore in an unstable state, and will spontaneously change over to one of the two magnetized states of least energy as the temperature is lowered below 770°C. It should be emphasized that in this example of the ferromagnet, nature has not abandoned symmetry. The symmetry is merely hidden below the Curie temperature.

Another example of temperature-dependent breakdown of symmetry occurs in the phenomenon of superconductivity, which we shall describe later.

How does the idea of spontaneous symmetry breaking help the modern theoretical physicist arrive at a unified electroweak theory? To answer this question, we have to introduce a new symmetry into the picture developed in the previous section, the picture which led us to the $SU(2)_L \times U(1)$ group.

This new idea, which began to play an important role in field theory in the 1970s, in fact dates from much earlier. Known as 'gauge theory', it was first proposed in 1918 by Hermann Weyl in the context of measurement in curved space–time. As the word 'gauge' indicates, Weyl was con-

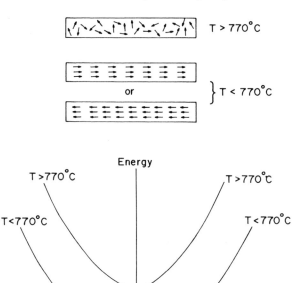

Fig. 4.6. At high temperatures the elementary magnets formed by spinning electrons in a bar of iron are randomly aligned, as shown by the arrows (top), and there is no net magnetic effect in the bar at the macroscopic level. Below 770°C temperature, however, as a magnetizing field is applied, the bar goes into a state of lower energy that happens to align all the electron magnets, as shown. There are two states of lower energy, depending on which side the magnetic north pole is. The curves indicate the energies at different alignments of the electron magnets.

cerned with the concept of measurement of lengths and angles at different points in curved space–time. When we make a statement like 'The length of a rod is 4 metres', what we really mean is that, taking a certain standard of length as 1 metre, the rod measures 4 such units. But when we take the rod from one point to another, or when we wish to compare lengths of two rods located at different points, are we really taking ratios of absolute lengths? No. For there is no guarantee that the standard of length (the gauge) itself has not changed in absolute terms from one point to another. In fact, as Weyl argued, such a change is unknowable by experiments. Hence the theoretical framework of any interaction should allow for arbitrary change of the measuring gauge from point to point. Einstein's general relativity did not allow for such arbitrary gauge changes. Weyl's motivation was his wish to construct a gravitation theory that did. We will return to this concept of gauge change in space–time geometry in Chapter 6.

In the terminology of the modern particle physicist, gauge symmetry in a physical theory implies its invariance under transformations which vary arbitrarily from point to point. This gauge-symmetry principle is thus a very stringent principle, and in its stringency lies its power. For the requirement of gauge invariance narrows down considerably the choice of a mathematical structure for a physical theory.

Maxwell's electromagnetic theory satisfies this requirement. This can be seen in the way electromagnetic fields are defined and in how they interact with charged particles. Let us take the wave function defining the electron, ψ_e. We have already introduced the notion of phase invariance; the transformation $\psi_e \rightarrow \psi_e \exp(i\theta)$ does not change the probability density of finding the electron in the space around any given point P.

But suppose that the phase transformation is made to vary from point to point. The phase angle θ then becomes a function of space–time. Would such a variation alter the structure of QED? Well, to begin with, the wave function ψ_e is supposed to satisfy a differential equation (the so-called wave equation) which tells us how it varies from one point in space–time to another. The introduction of θ brings in θ-dependent

terms which suggest at first sight that the physical structure of
the theory has in fact changed.

However, a closer look alters this conclusion. For the wave
equation satisfied by ψ_e includes a term which describes how
the electron, of charge $-e$, interacts with an electric field
described by a potential, $\mathbf{A}$. The potential vector, $\mathbf{A}$, has four
components, like any other vector in space–time. If our phase
transformation of ψ_e is accompanied by a corresponding
change in $\mathbf{A}$ given by

$$\mathbf{A} \to \mathbf{A} - \frac{1}{e}\nabla\theta$$

(here $\nabla\theta$ is the 4-dimensional gradient of θ), then it is found
that the wave equation for ψ_e is unchanged.

So far, so good! But by altering $\mathbf{A}$, have we altered the
electromagnetic fields? The beauty of Maxwell's equations is
that they are unaffected by this transformation! Thus QED is
unchanged under the above gauge symmetry. We also find that
by a suitable choice of θ, we can give the wave equation
satisfied by $\mathbf{A}$ in a simple form:

$$\mathbf{A} \equiv \frac{1}{c^2}\frac{\partial^2 \mathbf{A}}{\partial t^2} - \frac{\partial^2 \mathbf{A}}{\partial x^2} - \frac{\partial^2 \mathbf{A}}{\partial y^2} - \frac{\partial^2 \mathbf{A}}{\partial z^2} = 0,$$

where x, y, z are rectangular Cartesian coordinates for space,
and t is the time. This wave equation formally describes a
massless particle of spin 1. Known as the 'gauge boson', it is
none other than the photon, γ.

We will now restate what has been done above in a way
which illustrates the power of the gauge principle. Suppose
that a physicist had no knowledge of Maxwell's equations to
begin with. Then he would write the wave equation of ψ_e
without the interaction term. He would then discover that a
phase transformation of ψ_e in which the phase angle varies
from point to point alters the structure of this wave equation
by introducing an extra term—Q, say. If he now insists on
gauge symmetry—that is, that the equation be unaltered by
the phase transformation of ψ_e—he will be forced to invent an
additional term, E, which should have been included in his

original equation for ψ_e. The phase change of E should be such as to cancel exactly the unwanted extra term Q of the wave equation.

An analogy may help. Suppose an individual finds that an increase in his income has raised his income-tax bracket. If he wants to avoid paying higher taxes, he has to make suitable investments to qualify for deductions which will bring him back into the original tax bracket. In this way he discovers that savings which qualify for tax deduction exist in the economy.

Just as a tax accountant in the above example can work out exactly how such investments should be made, so the theoretical physicist would be able to calculate the type of additional term, E, needed to effect the required compensation. The answer he would come up with is that E describes an interaction of the electron with a vector field. In other words, with the requirement of gauge invariance, the physicist would be able to discover that a vector field **A** of zero mass, say, interacting with the electric charge should exist in nature.

The fact that in the above case θ is made to vary from point to point is often stated thus: the transformation $\psi_e \rightarrow \exp(i\theta) \cdot \psi_e$ is local rather than global. (In a global phase transformation θ would be the same at all points.) By going over from the global to the local version of the U(1) group, we have gauged the group. And, as we discovered in the above example, the gauging of the U(1) group leads us to the electromagnetic potential vector fields, **A**, and to QED.

However, the electromagnetic theory was known long before the gauge principle, and its 'rediscovery' in the above fashion is not by itself a great achievement. What is important about the above example is that the same game of gauging groups can be played with new groups, where the underlying dynamical field theory is still to be discovered.

For example, if we take the SU(2)$_L$ × U(1) group, we have to replace $\exp(i\theta)$ above by something like

$$\exp i(\theta_1 \sigma_1 + \theta_2 \sigma_2 + \theta_3 \sigma_3) \times \exp(i\theta).$$

If we insist on gauge symmetry, we have to make θ_1, θ_2, θ_3, and θ functions of position in space–time. As in the example of the $U(1)$ group, the changes in each of the four functions lead us to corresponding changes in a vector potential. Thus we have to begin with the four gauge bosons described by these potentials, all of them massless. There are, of course, complications which were not present in QED alone, such as, for example, the non-abelian nature of the $SU(2)_L$ group. But it can be shown that the resulting theory is renormalizable like QED. This was first shown by Gerhard t'Hooft in 1971 and is now seen to be a general feature of such gauge theories.

Although this suggests that we are on the right track in the search for an electroweak theory, there is still one important hurdle to cross. The four gauge bosons above are all massless, whereas we want only one of them, γ, to be massless, the other three, W^+, W^- and W^0, being massive. How can we achieve this without giving up the important dividend of renormalizability which came from the masslessness of all gauge bosons? This is where the notion of spontaneous breakdown of symmetry helps.

We introduced the concept of symmetry breakdown with the example of a ferromagnet. The phenomenon of ferromagnetism is observed on the macroscopic scale, although its origin lies in the microscopic feature of spin alignment. This macroscopic effect is seen because of *coherence* on the part of a large number of microscopic systems—in this case, the alignment of spins of electrons and iron nuclei all in the same direction. A similar situation arises in the phenomenon of superconductivity. It is found that some substances become superconducting *below* a certain critical temperature. In this case what happens is shown in Fig. 4.7. The superconductor does not allow any magnetic field to remain inside: it ejects the magnetic field lines from the interior to the periphery. This macroscopic behaviour is again due to coherence on the part of the microscopic components of the superconducting material.

The lack of penetration of the magnetic field into a superconductor is effectively equivalent to the electromagnetic field quantum acquiring a non-zero mass. For the acquisition of mass by the quantum reduces the range of its interaction (see

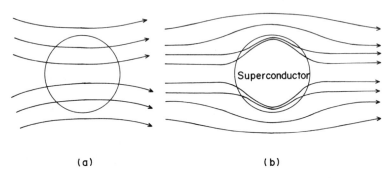

Fig. 4.7. A substance will normally permit magnetic lines of force to pass through it (a). Below a critical temperature, however, it may eject the lines totally and become a superconductor (b).

p. 118); in this case we say that the coherent behaviour of the microscopic components of the superconductor results in the photon acquiring a non-zero mass.

The microscopic components in this case are the so-called Cooper pairs of electrons (after L. N. Cooper who proposed this concept in 1956). An electron by itself is a fermion, but a pair of them behaves like a boson. How can it happen that two electrons form a unit? Normally, electrostatic repulsion will not allow such a pair to exist. In certain solids, however, the overall electric field of the oppositely charged ions can override this repulsion, and so Cooper pairs are formed. Being bosons, there is no limit on their population in any given state. (Recall the differences between Fermi–Dirac and Bose–Einstein statistics.) That is how it is possible to generate a macroscopic effect through coherence of a large number of Cooper pairs in the same state.

The purpose of our excursion through the byways of ferromagnetism and superconductivity has been to illustrate the trick which is used to resolve the problem of the electroweak theory. To this end, a boson field is introduced with two complex components $\phi_1 + i\phi_2$, and $\phi_3 + i\phi_4$, the field being linked to the four gauge bosons A_1, A_2, A_3, and A for the $SU(2)_L \times U(1)$ group. This two-component field has the symmetry of the SU(2) group, and its bosons are called 'Higgs bosons' (after Peter Higgs who, in 1964, first used this trick in particle physics). The coherence effect arises in the vacuum of this field. The mathematics of the process is rather involved, but its physical content can be understood in the following way.

As we have already mentioned, the vacuum in quantum field theory is non-trivial. Rather than it being a state in which there is nothing, it is more realistic to describe it as a state of lowest energy for the field system. When this criterion is applied to the Higgs field, it is found that the lowest-energy state is not necessarily one in which all ϕs are zero. Depending on the constants which go into the expression for the energy of the ϕ-field, the 'true vacuum' corresponds to a state wherein $\phi_1 = 0$, $\phi_2 = 0$, but $\phi_3^2 + \phi_4^2 = $ constant > 0. The state of $\phi_1 = \phi_2 = \phi_3 = \phi_4 = 0$ in fact has a larger energy than other 'nearby' states, and is consequently unstable. It is sometimes called a 'false vacuum'.

A familiar picture often used to describe the energy of the ϕ-field and illustrated in Fig. 4.8 is that of the bottom of a glass bottle. There is an elevation at the centre of the base which corresponds to the false vacuum, while the depression along the circumference corresponds to the true vacuum. Notice, however, that any point on the circle given by $\phi_3^2 + \phi_4^2$ = a constant is admissible as a vacuum state. A specific choice like $\phi_4 = 0$, $\phi_3 > 0$ is equivalent to the spontaneous breakdown of the rotational symmetry implied by the SU(2) group which operates on the ϕ-field.

This is when the coherence effect comes into play. The specific choice of vacuum value of the ϕ-field results in the three gauge bosons of SU(2)$_L$ acquiring non-zero masses, while the fourth gauge boson of U(1) stays massless—exactly what was required!

The masses acquired by the three bosons are related to the so-called coupling constants of the electroweak theory. Recall that when we gauged the U(1) group and 'deduced' the existence of the electromagnetic potential, we also allowed for an arbitrary coupling constant—namely, the electric charge, e,

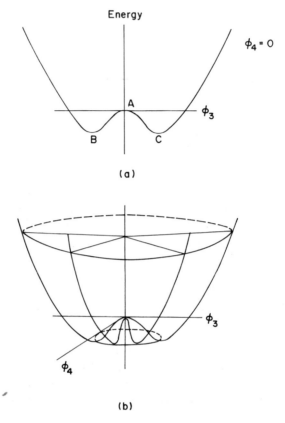

Fig. 4.8. The true and false vacua of the Higgs field, ϕ, are illustrated in (a) by the double-humped curve for the energy of the ϕ-field. The false vacuum is located at $\phi_3 = 0$, at A. The two states of true vaccum are located at points B and C, and the energy at these points is lower than the energy at A. If we rotate the curve around the axis (b), we get all the true vacuum states like B and C as points lying on the circle $\phi_3^2 + \phi_4^2 = $ constant parallel to the ϕ_3, ϕ_4 plane.

whose magnitude could not be deduced by the gauge principle. Likewise, when gauging the $SU(2)_L \times U(1)$ group, we need two coupling constants—g_1, g_2, say—whose magnitudes are not determined from the gauge theory. They have to be calculated from experimentally measured quantities. Once they are so determined, the electroweak theory predicts the masses of the three gauge bosons.

The mathematical gymnastics which led to this conclusion are considerably more intricate than those in Maxwell's equations or in Einstein's general relativity. The coherence effect produced by the vacuum is not directly observable: its consequence is to be seen in the mediating bosons of the weak interaction acquiring non-zero masses. Experiments with the high-energy accelerators at CERN in Geneva and the Fermilab in Chicago produced several instances in the 1970s of processes which implied the existence of the neutral weak boson. These experiments lent credibility to the electroweak theory of Weinberg, Salam, and Glashow, and made gauge theories popular among particle physicists.

The neutral weak boson in this electroweak theory is the realistic version of the expected neutral boson W^0. It is called the Z-boson, while the charged bosons are called the W-bosons. The masses of these bosons could not be predicted *ab initio*, but they could be deduced from the experimentally determined parameters of the theory. In 1983 Carlo Rubbia and his colleagues found from their studies of collisions between protons and antiprotons (in the proton–antiproton collider at CERN) certain events that gave proof of the existence of the W- and Z-bosons and also led to measurements of their masses. Their results were as follows: $M_W = 81 \pm 2$ GeV, and $M_Z = 93 \pm 2$ GeV. These are in very good agreement with the theoretical estimates.

In the last analysis, experimental verification of this kind generates confidence that the theory, however complex it sounds, is on the right track. It is this confidence which guides the particle physicist towards the more ambitious goal of unifying *all* physical interactions.

Grand Unified Theories (GUTs)

The reader may wonder, in spite of this sketchy description (or perhaps because of it!) whether real unification has been achieved in the $SU(2)_L \times U(1)$ electroweak theory. As we have seen, the $SU(2)_L$ group came from the weak processes and the $U(1)$ from the electromagnetic ones. By writing a multiplication sign between the two groups, we put the two processes together. But this is *not* unification in the sense that electricity and magnetism were unified by Maxwell. This is apparent from the mathematical structure of the electroweak theory: it starts by ascribing one coupling strength (g_2) to the gauge transformations described by the $SU(2)_L$ group and another (g_1) to the gauge transformations described by the $U(1)$ group. The ratio g_2/g_1 is measured experimentally, rather than determined theoretically. In this sense the theory is incomplete. The possibility of determining this ratio theoretically lies in successfully embedding this combination of groups, $SU(2)_L \times U(1)$, into a larger *single* group.

But at the same time why not look for a triple unification which would also bring the strong interactions into the fold? As we shall see shortly, the strong processes seem to be governed by another special unitary group, the $SU(3)$ group; so what we need is a group which contains the combination $SU(3) \times SU(2)_L \times U(1)$.

All the particles known today seem to fall into two groups: hadrons and leptons, depending on whether or not they take part in the strong interaction. Thus neutrons, protons, and mesons which take part in the strong interaction are classified as hadrons. For them, the strong forces naturally dominate over the electroweak ones (although, as we mentioned above, the latter forces cannot be ignored altogether). This difference between hadrons and leptons, which do not take part in the strong interaction, must somehow be reflected in their internal structure.

The pioneering work of Murray Gellmann and George Zweig in the early 1960s gave rise to the first clues to this mystery. Again we will jump across the historical details to our present-day understanding. To put it in a nutshell: today, hadrons are believed to be made up of quarks, which are not constituents of leptons.

Quarks, usually denoted by the letter 'q', come in several types, colours, and flavours. These three abstract properties are needed to explain the proliferation of hadrons revealed by various experiments with high-energy accelerators. Let me introduce these properties beginning with familiar examples of nucleons—that is, the protons and neutrons which make up atomic nuclei. According to the meson theory of Yukawa, first proposed in 1935, nucleons 'feel' the strong force through the exchange of bosons called π-mesons, or pions:

$$n \rightarrow p + \pi^-, \quad p \rightarrow n + \pi^+,$$

$$n \rightarrow n + \pi^0, \quad p \rightarrow p + \pi^0,$$

and so on. The analogy between the W-bosons which mediate the weak interaction and the above poins is clear; in fact, it was the above nuclear processes which inspired the intermediate-boson picture for the weak interaction.

However, unlike the e, ν_e, $W^\pm$, Z family, the p, n, $\pi^\pm$, π^0 group is understood in terms of quarks. Nucleons are made up of groups of three quarks, whereas pions are made up of pairs of quarks and antiquarks:

$$n = q_u q_d q_d, \quad p = q_u q_u q_d$$

$$\pi^+ = q_u \bar{q}_d, \quad \pi^- = q_d \bar{q}_u, \quad \pi^0 = q_u \bar{q}_u + q_d \bar{q}_d$$

The subscripts u and d stand for 'up' and 'down', the two types of quarks. The up quark has an electric charge of $+2/3$, the down quark a charge of $-1/3$ in units in which the electric charge of a proton is $+1$, that of an electron -1. The quarks are fermions with spin $\pm 1/2$. The nuclear transformations involving n, p, and π can now be understood in terms of changes in quark structure.

Just as the lepton number, L, was found to be conserved in the electroweak processes, so we can define a baryon number, B, which is conserved in the strong interaction. Quarks are given a baryon number, $B = 1/3$, whereas antiquarks have $B = -1/3$. Thus, in the above example the neutron and the proton have $B = 1$, whereas the pions have $B = 0$. For the sake of completeness, we set $B = 0$ for leptons and $L = 0$ for quarks.

Just as there are three flavours of lepton pairs, it seems that there are three corresponding flavours of quark pairs—combinations of one up and one down quark. The correspondence goes as follows:

$$\begin{pmatrix} e^- \\ \nu_e \end{pmatrix} \leftrightarrow \begin{pmatrix} q_u \\ q_d \end{pmatrix}, \quad \begin{pmatrix} \mu^- \\ \nu_\mu \end{pmatrix} \leftrightarrow \begin{pmatrix} q_c \\ q_s \end{pmatrix}, \quad \begin{pmatrix} \tau^- \\ \nu_\tau \end{pmatrix} \leftrightarrow \begin{pmatrix} q_t \\ q_b \end{pmatrix}.$$

The subscripts c, s, t, and b stand respectively for 'charm', 'strange', 'truth' (or 'top'), and 'beauty' (or 'bottom'). Just as the muon is a look-alike of the electron except for its heavier mass, so the similarity but extramassiveness of the K-meson ($q_u \bar{q}_s$) compared with the pion ($q_u \bar{q}_d$) suggests that the strange quark is more massive than the down quark. Why such look-alikes are allowed in nature is still a mystery.

In addition to flavours, quarks also possess another distinctive abstract property called 'colour'. Three colours are assigned to quarks—red (R), blue (B), and green (G)—and three corresponding anticolours to antiquarks—cyan ($\bar{R}$), yellow ($\bar{B}$), and magenta ($\bar{G}$). These colours have nothing to do with the colours we perceive in macroscopic objects, of course, but their identification with the abstract distinctive property of quarks helps us to visualize quark behaviour. For example, the choice of colours and anticolours is determined by the fact that lights of these pairs of colours combine to give white light. (See Fig. 4.9.)

Why introduce this feature at all? The need is as follows. Since quarks are regarded as fermions, restrictions are imposed by the Pauli exclusion principle (see p. 106) on combinations containing identical quarks. For example, because a quark has only two spin states, $\pm 1/2$, a combination like $q_u q_u q_u$ would be forbidden by the Pauli principle. For this would mean that of the three quarks at least two would have the same spin, which is not allowed. Yet a baryon with this

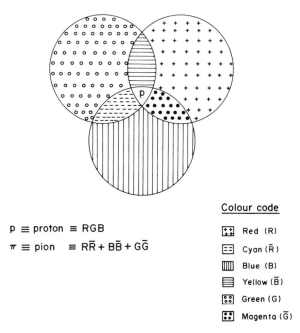

p ≡ proton ≡ RGB

$\pi \equiv$ pion $\equiv$ R$\bar{\text{R}}$ + B$\bar{\text{B}}$ + G$\bar{\text{G}}$

Colour code

⬒ Red (R)

⬓ Cyan (R̄)

▥ Blue (B)

▤ Yellow (B̄)

⬙ Green (G)

⬗ Magenta (Ḡ)

Fig. 4.9. How coloured quarks combine to produce the particles observed in nature.

combination, Δ^{++}, containing two units of electric charge, is found to exist. So the three quarks in $q_u q_u q_u$ must have another physical property (apart from spin) to distinguish them. The assignment of different colours to the three quarks is the solution.

Not all colour combinations are allowed, however! The hadrons which actually exist in nature must have combinations of colour such that the overall effect is 'white'. Thus the triplet $q_u q_u q_u$ permitted as a combination for protons must be made of equal mixtures of R, B, and G quarks. Similarly, mesons are equal mixtures of R$\bar{\text{R}}$, B$\bar{\text{B}}$, and G$\bar{\text{G}}$.

This colour combination rule has profound implications for hadron physics. To begin with, we see that single quarks or combinations like $q\bar{q}$ and $\bar{q}q$ containing fractional charges cannot exist. Obviously colour has some fundamental role to

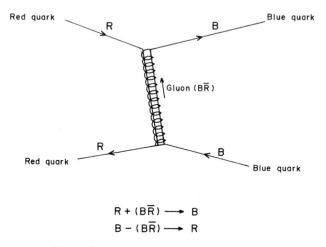

$$R + (B\overline{R}) \longrightarrow B$$
$$B - (B\overline{R}) \longrightarrow R$$

Fig. 4.10. How quarks change colours by exchanging gluons.

play in the way quarks interact with each other. In fact, it turns out that colour plays a role analogous to that of electric charge in QED.

This role can be visualized by thinking of two quarks as interacting through the exchange of a boson, just as two electrons in QED interact by exchanging a photon. These bosons are called 'gluons'. Unlike the photon, which itself is chargeless, gluons themselves carry colour combinations. Since these combinations are non-white, gluons cannot exist as 'real' particles; nevertheless, they play an important role in bringing about colour changes in quarks. For example, as shown in Fig. 4.10, if a red quark meets a blue quark coming from the opposite direction, the two can change colour by exchanging a gluon of colour B$\overline{R}$. This gluon added to the red quark changes it into a blue quark. By combining any of the three colours with any of the three anticolours, we would appear to have nine such colour combinations for gluons. However, not all nine can change colours in independent ways. For the combination R$\overline{R}$ + B$\overline{B}$ + G$\overline{G}$ containing equal mixtures of R, B, G, and their anticolours is white, and thus represents a real particle (a meson). Therefore, it cannot bring about a change of colours. It turns out that this is the only combination of this type, so that effectively we have *eight* different types of gluons.

The reader may have noticed the analogy between the present scenario and the SU(2) picture we developed for the electron–neutrino combination. Now we have three colour states in the wave function, ψ:

$$\psi = \begin{bmatrix} \psi_R \\ \psi_B \\ \psi_G \end{bmatrix}$$

Following the procedure of page 116 for this triplet, we arrive at the SU(3) group. The Hermitian matrix, H, is now a 3×3 matrix with zero trace. It therefore has eight real parameters, $\theta_1, \ldots \theta_8$ analogous to the three quantities θ_1, θ_2, and θ_3 for SU(2). Just as we had three bosons for SU(2), we now have eight for SU(3). These are precisely the eight gluons mentioned above. This rule can easily be generalized: a theory using an SU(n) group will have $n^2 - 1$ mediating bosons.

A theory of strong interactions between quarks and gluons can now be developed along the lines of QED, as an SU(3) gauge theory with eight 'colour charges'. Not surprisingly, it is called 'quantum chromodynamics' (QCD). For reasons discussed earlier, QCD is renormalizable. Because, like the SU(2) group, the SU(3) group is non-abelian, the gluons (like their 'weak' counterparts) interact directly with one another. (To indicate that this group relates to colour changes, it is often written as SU(3)$_C$, as we shall do henceforth.) Such direct couplings have also led to some differences between QED and QCD. A major difference relates to how powerful the interactions remain as one approaches their respective source particles.

We have already seen in the case of QED that there is no such thing as a single isolated electron. An electron, E, in a vacuum is really surrounded by particle–antiparticle pairs (e^-, e^+) which are continually being created and annihilated. Now because of the Coulomb force, the electron in question will tend to attract the oppositely charged positron (e^+) of the pair and repel the like, charged electron (e^-). The outcome is that a typical vacuum pair will be preferentially oriented with respect to E, with e^+ closer and e^- further away from it. Since the electron is surrounded by a cloud of such oriented pairs, its effective electric charge is *screened* by the cloud. So the closer to E that one penetrates in this cloud, the less will be the effect of the screening. In other words, if we bombard E with a probe charge, Q, of increasing energy, we find that the higher the energy of Q, the closer it gets to E, and the larger the charge of E as measured by Q. To put it in technical jargon, the effective strength of the electromagnetic interaction is found to increase with the energy of the participating electric charges.

Now the reverse is true of the strength of the colour interaction between quarks, as shown in Fig. 4.11. Because of the gluon–gluon interaction, the effective colour charge of a quark is found to increase as one moves *away* from it. A typical colour, say blue, attracts like colour, so that a cloud surrounding a blue quark would tend to enhance the quark's blueness as measured from outside. The closer one approaches the quark, the less intense one finds its blue colour. Again, in technical jargon, the effective strength of QCD decreases at high energy. Going to the limit of infinite energy (that is, getting very close to the quark), one would find that the strength of the interaction is zero! Quarks, then, behave essentially as free particles. This deduction is of course based on extrapolating the behaviour of QCD at moderately high energy, and is called 'asymptotic freedom'. As we have just seen, it is the antiscreening effect of gluons which is responsible for asymptotic freedom.

In this opposite behaviour of QED and QCD at high energies lies the hope of their unification. For at moderate energies at which the participating particles all move with speeds slow compared to the speed of light, the strength of QED is about 1 per cent of the strength of QCD. At high enough energy, however, the two interactions may be of comparable strength. This is where unification may be both achieved and perceived.

To give an analogy from electromagnetism: a slow-moving

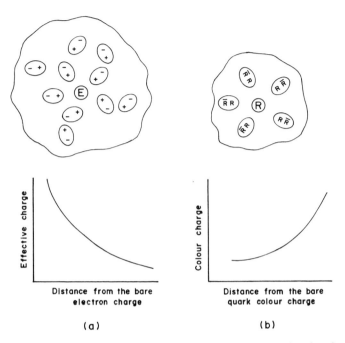

Fig. 4.11. In QED (a), the electron charge is screened by the cloud of oppositely oriented electron–positron pairs surrounding it. In QCD (b), the cloud of gluons antiscreens the intensity of the colour of the quark at its centre.

electric charge feels the effect of an electric field, but very little of the effect of a magnetic field. However, if it moves at speeds approaching the speed of light, it is subject to comparable forces from the electric and magnetic fields. Indeed, the speed of light is the key speed which brings about a unification of electricity and magnetism.

Let us now return to our previous discussion and consider the possibility of unifying QCD with the electroweak theory by looking for a group, G, which contains the $SU(3)_C \times SU(2)_L \times U(1)$ group. What we said above suggests that the effect of G, if it exists, will be perceived at very high energies. How high?

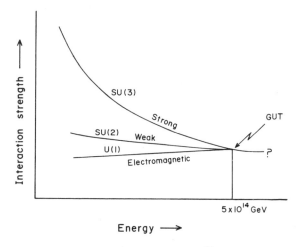

Fig. 4.12. At energy of the order of 5×10^{14} GeV the strengths of the electromagnetic interaction, the weak interaction, and the colour interaction become comparable. This is where grand unification is expected to occur.

Without going into technical details, we can rephrase the question in the following way. Suppose that in the combination above, the three groups $U(1)$, $SU(2)_L$, and $SU(3)_C$ occur with coupling constants g_1, g_2, and g_3 respectively. For the electromagnetic case g_1 is just the magnitude of the electronic charge, e, which we have taken as our unit. We can construct a dimensionless quantity from g_1 called the 'fine structure constant':

$$\alpha_1 = \frac{e^2}{\hbar c}$$

For the moderate energies at which QED is studied in the laboratory, $\alpha_1 \cong 1/137$. We may similarly define α_2 and α_3 in dimensionless form from g_2 and g_3.

From the screening/antiscreening property of the vacuum, it is clear that these coupling strengths, α_1, α_2, and α_3, will change at high energies: α_1 will increase with energy, whereas α_2 and α_3 will decrease as the energy is increased. Fig. 4.12

illustrates this schematically. Present calculations indicate that α_1, α_2, and α_3 approach comparable values at an energy of about 5×10^{14} GeV. This energy is more than five million million times the energy associated with the W- and Z-bosons. We may associate this energy with the masses of a family of bosons called 'X-bosons', which play the key role in unification. Their mass, $M_x \cong 5 \times 10^{14}$ GeV.

So far we have not said what group, G, might be involved in this 'grand' unification of the three interactions. (The choice of the word 'grand' is due to Sheldon Glashow. That words like 'grand' and 'super' should have crept into the matter-of-fact field of science is a reflection of how media-oriented science has become.) H. Georgi and S. Glashow have shown that the smallest group to contain the triple combination $SU(3)_C \times SU(2)_L \times U(1)$ is the $SU(5)$ group. This describes rotations of a wave function of five components, which we may loosely identify with the three colour quarks and the two leptons all belonging to the same flavour. We can see that there will be $5^2 - 1 = 24$ mediating bosons. Of these we already have the 8 gluons, the $W^{\pm}$-bosons, the Z-boson, and the photon, γ. So we need 12 additional bosons, which are the superheavy X-bosons conjectured above.

There are, of course, larger groups than the $SU(5)$ group which also contain the triple product $SU(3)_C \times SU(2)_L \times U(1)$. By the usual appeals to simplicity and economy, however, we will turn to those only if the simplest possibility fails to conform to the experimental checks. So, what do the experiments tell us?

GUTs and Experiments

Can GUTs be verified by experiment? The present reach of high-energy accelerators is at most 10^5 GeV, which falls far short of the unification energy of $\sim 10^{15}$ GeV. However, with the electroweak theory, some low-energy experiments indicated effects associated with the $W^{\pm}$- and Z-bosons long before accelerators could reach energies as high as 100 GeV. One familiar example is that of the muon decay $\mu \rightarrow e^- + \bar{\nu}_e$

+ ν_μ which proceeds via the mediation of the W-boson. Can we similarly see some effect of an X-boson without having to reach the fantastically high energy of 10^{15} GeV?

The answer is yes. In the SU(5) framework, for example, X-bosons would play a role in converting quarks to leptons, and vice versa. Since quarks make up baryons, we therefore expect to see violations of baryon number as well as lepton number. Can such conversions be observed in laboratory experiments? The SU(5) theory predicts that even the stable proton should decay, one mode of decay being $p \rightarrow e^+ + \pi^0$. The lifetime of the proton according to the theory comes out to be a few times 10^{31} years! Notice that in such decays the number $B - L$ is still conserved.

How can we ever expect to observe events whose characteristic time is so huge? The situation is not hopeless if we consider the example of radioactive decays, however. Decays of nuclei like neodymium and tellurium are observed, even though their lifetimes are measured as longer than 10^{15} and 10^{21} years respectively. The main point to note is that a particle like a proton can decay spontaneously *any time*, although the chance of its doing so in a year is as low as 1 in 10^{31}. The smallness of the decay probability can be compensated by using large numbers of particles. Protons are certainly very abundant, and 160 tons of ordinary matter contains about 10^{32} protons. Thus, in such an amount, we may expect about 1–10 protons to decay in a year on the average. If we know what the decay products are going to be, we can set up detectors to look for them.

Of course looking for a proton decay event is more difficult than the proverbial search for a needle in a haystack. The smallness of the number of decays expected makes it hard to distinguish them from other 'background' events induced by cosmic rays, neutrinos, and so forth. To minimize the background, it helps to conduct such experiments deep underground.

The pioneering experiment in this field was set up in the late 1970s in a gold mine of Kolar Gold Fields in South India under the joint sponsorship of the universities of Tokyo and

Osaka in Japan, and the Tata Institute of Fundamental Research in India. The protons are in a pile of iron of mass ~140 tons. The detectors are proportional counter modules filled with a mixture of argon and methane. The decay modes and their relative frequency depend on the underlying theory, but in general the hope is to detect muons, pions, or other mesons. To be sure that the observed particle tracks are due to proton decay, it is necessary that they originate in the *interior* of the pile. (Background events tend to cause tracks from the surface to the surface.)

Since its inception this experiment has yielded four or five likely candidate events of proton decay. If they are genuine, the proton lifetime works out at no less than 2×10^{31} years. Other experiments set up elsewhere in the world (there are about ten of them, some using water as the pile of protons) suffer from larger numbers of background events, and so far they have not yielded any tangible results, although the null result of proton decay in a water reservoir in the Irvine–Michigan–Brookhaven experiment seems to imply a proton lifetime longer than 10^{32} years. The results from various detectors are summarized in Table 4.1.

This tenacity on the part of the proton puts severe constraints on the SU(5) theory, and may well rule it out altogether. But even if it is ruled out, particle physicists have other GUTs to fall back on. They can also invoke so-called supersymmetric theories, which we shall refer to briefly in the next chapter.

The sceptic and the purist will not be satisfied even with positive results from proton decay experiments, however. Such results would go as far as presenting prima-facie evidence for grand unification, but they would not vindicate any particular GUT model. For that, dynamical experiments would be needed.

Recall that the electroweak theory gained a lot of credibility from high-energy accelerator experiments culminating in measurements of the masses of the W- and Z-bosons. Even Maxwell's electromagnetic theory only became really established after Hertz's demonstration of electromagnetic waves in the

Table 4.1 Summary of Results from Proton Decay Experiments

Experiment	Location	Collaborators	Exposure (Kilotons × years)	Confined events	Candidate events
KGF	Kolar Gold Fields, India	TIFR, Bombay, and universities of Tokyo and Osaka City	0.23	20	4
NUSEX	Mt Blanc, Italy	CERN–Milano–Frascati–Turin	0.22	22	2
IMB	Ohio, USA	Irvine–Michigan–Brookhaven	3.80	401	—
HPW	Park City, USA	Harvard–Purdue–Wisconsin	0.38	5	2
Kamioka	Kamiokande, Japan	Universities of Njigata and Tokyo–KEK	0.66	89	3
Fréjus	Fréjus, France	Aachen–Orsay–Palaiseau–Saclay–Wuppertal	0.09	9	—

laboratory. But the hope of verifying GUTs by high-energy experiments in the laboratory is dashed by the very large particle energies required ($\sim 10^{15}$ GeV).

Are GUTs doomed to be mere speculative exercises then? Science has always insisted that knowledge be based on experimentally verified facts, and a theory which cannot be verified experimentally or observationally is not taken seriously by scientists. Indeed, such would have been the fate of GUTs but for the timely appearance on the scene of big bang cosmology.

As we have seen in earlier chapters, the hot big bang origin of the universe indicates conditions very early on under which particles of arbitrarily high energies could exist. The temperature–time relationship $T \propto t^{-1/2}$ in the early universe tells us that as time, t, approaches zero (that is, as we approach the epoch of the big bang), the temperature T shoots to infinity. Since the temperature reflects the average energy of the particles present, we see that the hot universe can act as the high-energy accelerator required for testing GUTs.

We began this chapter by highlighting the cosmologist's problems, and how he needs input from particle physics. In GUTs we see hopes of solving these problems. At the same time we now find that the particle physicist needs help from the cosmologist: for the only situation in which GUTs could probably have operated to the full was that of the very early universe. Not surprisingly, therefore, co-operation has developed between cosmologists and particle physicists, a co-operation which has yielded some exciting ideas. We will discuss these next.

5

The Very Early Universe

We have been engaged in *world-building*.... There is
little satisfaction to the builder in the mere assemblage
of selected material already possessing the properties
which will appear in the finished picture. Our desire is
to achieve the purpose with unselected material. In
the game of world-building we lose a point whenever
we have to ask for extraordinary material specially
prepared for the end in view.

A. S. Eddington

The proof of the pudding lies in the eating. Applied to science,
this saying demands factual verification of theoretical predictions. In Chapter 4 we saw that a possible, and perhaps the
only, verification of theoreticians' speculations about the unity
of physical forces may be found in how the universe behaved
shortly after the big bang. Investigations of the very early
universe which seek to match facts with speculations thus
comprise the most exciting frontier area between particle
physics and cosmology today.

How can we look for facts relating to the very early universe, to events which took place in fleeting moments—we will
see how fleeting, shortly!—some ten to fifteen billion years
ago? The method of direct observation of the distant parts of
the universe has its own limitations. It is true that because
light has a finite velocity, the farther away we look, the earlier
the epochs we observe. So what is wrong with the expectation
that a supertelescope of the future able to probe the universe
out to fifteen billion light-years will bring us direct information about the very early universe? The expectation cannot be
realized because of the cosmic radiation background. Recall
from Chapter 3 that at epochs prior to that of red-shift 1,000,

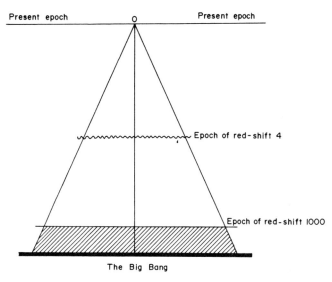

Present epoch   O   Present epoch

Epoch of red-shift 4

Epoch of red-shift 1000

The Big Bang

Fig. 5.1. The past light-cone of the present observer, O. The wavy line indicates approximately the distance (and epoch) out to which our present telescopes can probe the universe. The shaded area represents the epoch at which the radiation was so intense that no coherent observations would be possible at O. (The time axis is not drawn to scale.)

the universe was dense to radiation. Light photons bringing information about those earlier epochs get knocked about so much that no coherent message can ever reach the present-day observer. Fig. 5.1 illustrates this point.

There is another way in which the fact-finding mission may proceed, however. Whatever happened in the very early universe under the influence of a GUT (or any other physical theory) might have left permanent relics which are observable today. The currently observed microwave background and the abundances of light nuclei like deuterium and helium may be considered relics of what went on in the universe from ~ 1 second to $\sim 100,000$ years after the big bang. Are there other relics, which carry the signatures of even earlier epochs?

The Photon-to-Baryon Ratio

The physicists' way of approaching cosmology gives a clue regarding the answer to this question. In a big bang cosmology, the present state of the universe must be ultimately traceable to conditions in the very early epochs, through 'cause and effect' arguments of physical laws. So, by definition, the entire observable universe of today is a relic of the very early epochs. However, the cause and effect chain to the past may have several links, not all of which are relevant to the very early universe.

Take, for example, the state of matter in the universe. At present we see matter in solid, liquid, and gaseous states, made up of chemical elements or compounds. But these states came about relatively recently. Nuclei of elements like carbon, oxygen, and iron were synthesized in hot stellar interiors, from neutrons and protons much later than the big bang. But how and when were neutrons and protons made? For making even deuterium and helium in the early universe, we had to take for granted the existence of protons and neutrons. In short, we may look upon baryons (protons and neutrons) as important relics of the very early universe.

Side by side is the question of antimatter. Does the universe consist predominantly of matter? Do baryons vastly outnumber antibaryons? Since baryons and antibaryons tend to annihilate each other and produce radiation, we should be able to explain the relative abundances of these three relic species: baryons, antibaryons, and photons in the universe.

How well do we know the matter–antimatter composition of the universe? Our main, and often the only, source of information from distant parts of the universe is electromagnetic radiation, which behaves symmetrically with respect to matter and antimatter. In other words information regarding structural details, composition, and so on carried by visible light, X-rays, radio waves, and other forms of electromagnetic radiation cannot tell us if a distant quasar or a cluster of galaxies is made up of matter or antimatter. Somewhat nearer home we may be more confident, however. For example, cosmic rays in our galaxy and perhaps from some nearby galaxies tell us that they come from sources composed of *matter*. Extrapolating, we may extend this result to the cluster of which our galaxy is a member. Beyond that we cannot be so assertive, except to argue that since there is a large-scale structural homogeneity in the universe, this also implies a homogeneity with regard to the species of its contents.

Assuming that all particles in the universe consist of matter (implying a universe biased in favour of matter, despite the basic matter–antimatter symmetry) and that the radiation at present is predominantly in the form of microwaves, we already have (see p. 101) the ratio of number densities of photons and baryons as estimated *at present*:

$$\frac{N_\gamma}{N_B} = 4.57 \times 10^7 \, (\Omega_0 f_B h_0^2)^{-1} \left(\frac{T_0}{3}\right)^3.$$

Although some uncertainty persists in the observed values of the Hubble constant, h_0, the density parameter, Ω_0, and its baryonic fraction, f_B, as well as in the temperature of the microwave background, T_0, the above formula tells us that N_γ/N_B lies in the broad range of 10^8–10^{10}. (See Fig. 5.2.)

As noted earlier, both N_γ and N_B were higher in the past in inverse proportion to the cube of the cosmic expansion factor, S; so the ratio N_γ/N_B has remained unchanged since some very early epoch when the number of baryons was somehow fixed. This ratio is therefore a relic of the very early universe.

In the early 1970s, before GUTs really got under way, physicists argued in the following manner regarding N_γ/N_B. Suppose that 'initially' the cosmic brew had equal numbers of baryons and antibaryons. These would interact with each other and produce radiation thus: $B + \bar{B} \leftrightarrow \gamma + \gamma$. The double-headed arrow denotes that the reaction proceeded in both directions: just as baryon–antibaryon pairs annihilated to produce photons of radiation, so photons collided to produce baryon–antibaryon pairs. The strong interaction was expected to determine how fast this reaction would go either way, and

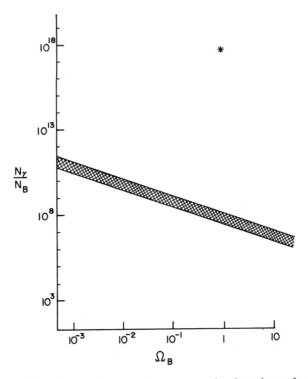

Fig. 5.2. The 'observed' photon-to-baryon ratio plotted as a function
of $\Omega_B = \Omega_0 f_B$. The shaded region is where the ratio lies for h_0 in the
range 0.5–1.0. Arguments from statistical mechanics lead to a much
higher value, typically that of the star at $\Omega_B = 1$.

in the initial phases it seemed that the reaction rate was *fast*
compared with the rate of expansion of the universe.

However, as time went on, both rates declined, the reaction
rate dropping more quickly than the expansion rate. Inevitably, there came an epoch when the former dipped below the
latter. Beyond this critical epoch we may assume that the
baryon–antibaryon (or photon–photon) collisions were too
infrequent to alter the numbers of baryons, antibaryons, and
photons significantly. Their numbers at that epoch were determined from considerations of thermodynamic equilibrium

along the lines described in Chapter 3. The result of the calculations[1] was somewhat startling:

$$\frac{N_\gamma}{N_B} = \frac{N_\gamma}{N_{\bar{B}}} \approx 5 \times 10^{17}.$$

In other words, there are far too few surviving baryons (and antibaryons) per photon in the universe: and the discrepancy between theory and observation is seven to nine orders of magnitude. Further, it seems necessary to assume that somehow the baryon population got well separated from the antibaryon population at some stage, so that in the currently observable universe only baryons are seen. Clearly 'new physics' is necessary to understand and resolve these problems.

GUTs held out the promise of a solution. If baryons can decay, so can they be produced, provided the energies of particles are high enough—above 10^{15} GeV, say—to make GUTs operational. How early on in the universe would one expect this to have been the case?

Assuming that the particles participating in the interaction were moving with speeds near that of light and were in thermo-dynamic equilibrium, with frequent collisions, the cosmo-logical equations of Chapter 3 can be applied to derive the time–temperature relationship in the following form:[2]

$$t_s = 2.4 \times 10^{-6} \, g^{-1/2} \, T_{GeV}^{-2},$$

where T_{GeV} is the temperature expressed in energy units of GeV. (To remind ourselves of these units, the following formula is helpful: 1 GeV $= 1.6 \times 10^{-3}$ ergs (energy) $= 1.16 \times 10^{13}$ K (temperature).) The quantity g, encountered in Chapter 3, is a count of the 'effective' number of spin degrees of freedom of all participating particles. Fig. 5.3 illustrates this relation. In the SU(5) framework g lies between 100 and 200 for three flavours of quarks and leptons.

Taking $g = 100$ to fix ideas, we find from Fig. 5.3 that an energy of 10^{15} GeV corresponds to $t = 2.4 \times 10^{-37}$ second. So a universal epoch as early as this is relevant to our problem of N_γ/N_B, if it is to be solved via GUTs. We will refer to this as the GUTs epoch.

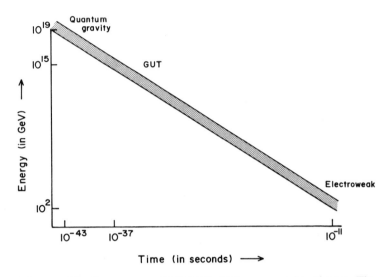

Fig. 5.3. The time–energy relationship in the very early universe. The broad band reflects uncertainty in the factor g.

However, more ingredients were needed to cook up the final solution. For example, GUTs not only allow creation of baryons; they likewise permit creation of antibaryons. If we are to emerge with an eventual excess of the former over the latter, we need some asymmetry between matter and anti-matter. A preponderance of baryon-creating reactions violates the symmetry of CP which we introduced in Chapter 4.

This is not all. For if we continue to assume that all the interacting particles were in thermodynamic equilibrium, then each reaction producing a net excess of baryons (or anti-baryons) will be countered by a reverse reaction, thereby eliminating any net baryon (or antibaryon) production. So we must assume that the stage at which excess baryons were being generated was one in which the participating particles were *not* in thermodynamic equilibrium.

Following these guide-lines, particle physicists like M. Yoshimura, S. Weinberg, F. Wilczek, J. Pati, and A. Salam were able to produce a plausible scenario for the very early universe, with a net photon-to-baryon ratio, N_γ/N_B, lying in the range 10^4–10^{12}. Although considerably broader in range than the observed value, this result had the prima-facie merit of including the observed range and explaining how it came about that the universe became antisymmetric in its matter–antimatter composition.

Such calculations are still of a tentative nature in that there are considerable uncertainties as to how GUTs operate. At best they may be considered a first step towards understanding the relic value of N_γ/N_B. While these calculations were being further improved and sharpened, however, the picture of the very early universe began to encounter fresh difficulties with other relics; and to surmount these problems new ideas emerged.

Some Problems of the Very Early Universe

Let us look briefly at these new problems before we consider the remedies proposed for them.

The horizon problem

Let us consider the past light-cone of a typical observer in the standard big bang model. This light-cone is, of course, the part of space–time accessible to light rays going *backwards in time*, starting from the observer's instantaneous position. To simplify the picture, ignore for the time being all the effects of curved space–time and imagine that this light-cone is drawn in the Minkowski space–time of special relativity. Retain, however, the important concept of the big bang, beginning at $t = 0$. Thus the space–time does not extend beyond $t = 0$.

Let the instant from which the light-cone is considered at the observer be T, and let the distance from the observer be r at any time t in the past, then we have along this cone $r = c(T - t)$, c being the speed of light. The light-cone can continue as far back in time as $t = 0$, by which time r has reached the value $R = cT$.

It is easy to interpret the past light-cone in the following way. Our observer at time T is not susceptible to physical influences from any object lying beyond a sphere of radius $R = cT$ centred on him. This is because no physical influence is supposed to travel faster than light, and there has simply not been enough time available for such an influence to complete the journey. This sphere of radius R is called the 'particle horizon' of the observer; it limits the zone of causal contact with that observer.

Of course, the size of the particle horizon grows with time, since more and more observers are able to communicate with the observer at $r = 0$. Conversely, as we go closer and closer to the big bang epoch the size of the horizon shrinks. This linear relationship between R and T holds even when we drop our simplifying assumption of Minkowski space–time and work out the past light-cone in terms of Friedmann models. At the GUTs epoch the horizon size was typically $\sim 10^{-26}$ cm.

The small size of the particle horizon close to the big bang naturally inhibits causal contact between regions well separated from one another. For example, if two observers A and B are separated by a distance exceeding $2R$, then at time T there can be no overlap between their particle horizons. In other words, there is no way in which either of them can know about the physical state of the other. Under these circumstances it would indeed be surprising if they happened to have the same physical state. (By way of analogy, imagine two pockets of human civilization well separated from each other's communication range, arriving independently at an identical lifestyle!)

Yet something like this is what is found in the actual universe. The large-scale isotropy of the microwave background observed today implies that regions which are now separated by as much as 10^{28} cm have the same physical conditions. Let us try to estimate the size of a region which today has linear dimensions of 10^{28} cm at the time when the universal temperature was 10^{15} GeV, on the assumption that this region has since expanded according to Friedmann models.

There is a simple way of making this estimate if we recall

that the radiation temperature falls essentially in inverse proportion to the universal scale factor (see p. 88). We know that the present temperature of the universe is 2.7 K—that is, about 2.5×10^{-13} GeV, from the conversion formula given earlier. The temperature drop from 10^{15} GeV to the present value thus involves a factor of 4×10^{27}. So this is the factor by which the linear scale of the universe has expanded since that very early GUTs epoch.

The above result leads us to the startling conclusion that at the GUTs epoch the homogeneity extended to a distance of 2.5 cm, some twenty-six orders of magnitude larger than the size of the particle horizon at that epoch! But there was no way that homogeneity on such an extended scale could be achieved by the normal mixing cum cause-and-effect method of physical processes. The cosmologist is therefore driven to the somewhat unpalatable assumption that the universe must have been *created* homogeneous. While this assumption is not an impossible one *per se*, it nevertheless implies that one of the striking features of the present universe—namely, its large-scale homogeneity—cannot be explained by known physics.

The horizon problem was in fact known to cosmologists long before GUTs. In the late 1960s, Charles Misner had argued that the horizon may be a feature of Friedmann models with their high degrees of symmetry, and that if one wants to *explain* the observed homogeneity and isotropy, one should not *start* with Friedmann models, but rather, *end* with them. Could it be that the universe started in a highly asymmetric way, but eventually became highly symmetric?

Misner proposed the so-called mixmaster model, in which the universe passed through a succession of randomly changing states of anisotropy. The anisotropy in any state was such that along one direction in space there was no particle horizon. Thus along that direction there was no limit on causal communication. By changing this direction at random, the hope was that mixing of the different parts of the universe would take place rapidly enough to achieve large-scale homogeneity. Detailed calculations by D. M. Chitre showed that this hope could not be realized, however; so the horizon problem remained unsolved.

The problem of monopoles

One consequence of GUTs is the prediction that magnetic monopoles should exist, with characteristic energy of 10^{16} GeV.

When Maxwell's equations of electricity and magnetism were formulated fully, it became apparent that isolated magnetic poles cannot exist in nature. A bar magnet behaves as if it has two poles, north and south, one at each end. Yet any attempt to isolate the two poles by chopping the magnet in two is doomed to failure; for each of the two pieces again has opposite poles. In this sense magnetism is different from electricity: we can, and indeed do, have isolated electric charges, of either kind, positive or negative.

This difference between electricity and magnetism shows up in many ways in nature. In our daily lives we have electric currents but not magnetic ones. Astronomers find large-scale magnetic fields, but not corresponding electric fields. If free monopoles had existed, they would have moved in such a way as to kill any large-scale magnetic fields.

However, monopole-like structures are possible, and indeed inevitable, in the framework of GUTs. Therefore one is forced to conclude that in the GUTs epoch such monopoles did exist. And, if they existed then, it also follows from theory that they should exist now; for once created, it is hard to think of a scenario for *destroying* a magnetic monopole. So monopoles once created are here to stay.

Let us estimate the mass density of magnetic monopoles on the assumption that they are relics of the GUTs epoch. Since the particle horizon at that epoch limited the range of physical interactions, we may legitimately assume that there was at least one monopole in a typical horizon-size region. At the GUTs epoch the horizon size was $\sim 10^{-26}$ cm, and consequently, the monopole number density was one per $(10^{-26}$ cm$)^3$. Since this region (as we saw earlier) has expanded by a linear factor of 4×10^{27}, this number density has now dropped to about 1.6×10^{-5} cm^{-3}. This is comparable to the present estimate of the baryon number density—namely, $\sim 10^{-6}$ per cm^3. Recall, however, that a typical monopole is 10^{16} times as massive as a typical baryon, which makes the mass density of the universe in the form of monopoles as high as $\sim 10^{-13}$ g cm^{-3} or 10^{16} times the closure density!

Cosmologically this is an absurd result. A universe as dense as this would not have continued to expand until the present epoch; it would have recollapsed to singularity long ago, as the $k = +1$ models do (see Chapter 2). This discrepancy shows that something is seriously adrift in this simple-minded application of particle physics to cosmology.

The problem of domain walls

Recall from Chapter 4 that spontaneous symmetry breaking forces a physical system to go into one of a possible range of states. In the case of the very early universe the states are those of the mediating Higgs bosons which, as we saw, play the key role in the symmetry-breaking process. Now in the GUTs breakdown of symmetry a typical Higgs boson may have to choose between two states described by a wave function s with values: $\phi = +A$ and $\phi = -A$, where A is some parameter from the theory.

Although it does not matter which value of ϕ is chosen, it could very well happen that there are domains in the universe where $\phi = +A$ which lie next to domains where $\phi = -A$. Whereas physical conditions in a typical domain are uniform, boundaries between domains of opposite value of ϕ are places where discontinuities are concentrated, much like the disturbed boundaries between hostile neighbouring countries. These are the domain walls, which carry huge concentrations of energy on their surfaces.

Such surface defects cannot be wiped out. Like monopoles, domain walls survive to the present day. If such a defect were to pass through our observable universe, its enormous energy concentration would be easily noticed. For example, there would be marked anisotropy in the dynamical behaviour of the universe in the direction of the domain wall. That the universe shows no such evidence of anisotropy is therefore a problem.

The problem of massive neutrinos

Are neutrinos particles with zero rest mass (like photons) which always travel at the speed of light? Until the 1970s the answer to this question was always yes. But because neutrinos interact only weakly with other matter, direct experimental verification of this statement was not (and still isn't) easy. So theoreticians have felt free to speculate about the consequences if the answer is no.

Cosmologists are motivated in this direction by growing evidence for the presence of so-called dark matter in galaxies and in clusters thereof. By definition, 'dark matter' is matter which cannot be detected by astronomical observations, which normally rely on electromagnetic radiation. Thus dark matter cannot be seen, but its presence may be inferred by its gravitational influence on the surrounding visible matter.

Let us take the example of a spiral galaxy like ours. Astronomical studies tell us that the bulk of its luminous matter (for example, in stars) is confined to a region not extending beyond a few thousand light-years from its centre, say. Let us assign a mass M to this luminous matter, and consider its gravitational influence on a chunk of matter circling it in the plane of the galaxy at a distance R well beyond the range of the visible mass concentration. Suppose we are able to measure the speed, V, with which this chunk is going round. Then Newton's law of motion equates the centripetal acceleration V^2/R of the piece with the gravitational force per unit mass acting on it:

$$\frac{V^2}{R} = \frac{GM}{R^2}.$$

This simple relation tells us that the speed of rotation of the piece drops off as the square root of its distance from the centre as we examine such pieces farther and farther out. This

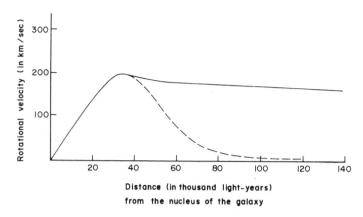

Fig. 5.4. The typical rotation curve of a galaxy extends flat outwards to distances of more than a hundred light-years, instead of dropping off along the dashed line, which is based on Keplerian orbits around visible matter in the galaxy.

relation in fact is none other than that obtained by Kepler in the seventeenth century for planets circling the sun.

In the 1970s radio-astronomical studies of the interiors of galaxies using the 21 cm wavelength began to reveal a somewhat different picture. The 'rotation curves' of galaxies, which show how V depends on R, did not taper as expected by the Keplerian law, as shown in Fig. 5.4, but they were almost flat for values of R considerably far out, extending to more than a hundred thousand light-years.

These 'flat' rotation curves may be interpreted in two ways. The more radical way is to assert that the law of gravitation does not follow the inverse square rule at galactic distances. The less disturbing way is to presuppose the existence of dark matter extending out to these large distances. For to maintain an undiminished value of V for large values of R, the above formula must have M/R constant—that is, M must represent not just a luminous mass concentrated at the centre, but also an invisible component extending farther out. The average density of matter, visible and invisible, is then thought to taper off as the inverse square of the distance from the galactic centre.

Observations of clusters of galaxies like the one shown in Fig. 5.5, while not definitive, appear to lead to the same conclusion. In this case one is looking for the distribution of velocities of galaxies as they move in one another's gravitational fields. Dynamical theory tells us that, given sufficient time, the distribution settles down to an equilibrium state with the property that the kinetic energy of motion is comparable to the gravitational potential energy. In practice the former can be estimated from velocity measurements of cluster members, while the latter is found from the mass estimates of galaxies *as actually visible*. Again, the kinetic energy turns out to be too high, and one needs to invoke dark matter to enhance the gravitational potential energy.

What could this dark matter be made of? Could it be in black holes? Could it be in 'jupiters'—that is, large planet-like masses which are not massive enough to initiate nuclear fusion in their interiors and shine like stars? These and other possibilities of a stellar nature lead us to consider dark matter as in the form of baryons. It is hard to accommodate this possibility, however, because it would increase the baryonic component of matter density and, as mentioned in Chapter 3, drastically reduce the primordial deuterium abundance in the universe. So baryonic dark matter is ruled out. In Chapter 6 we will return to dark matter alternatives under consideration today.

In 1972 Ramnath Cowsik and J. McClelland proposed that the dark matter might consist of massive neutrinos. In Chapter 3 we assumed that neutrinos were massless particles, and on that basis determined the temperature of their present-day background to be around 2 K. If it is assumed instead that neutrinos are massive, with rest-mass energy of a few electron volts, say, then the picture is altered. Instead of moving with the speed of light, the neutrinos would all have come to rest. And therefore their effect on the rate of expansion of the universe would be significant. For, though their masses may be small compared to the masses of baryons, their numbers would

Fig. 5.5. The Hercules cluster of galaxies. Photograph courtesy of National Optical Astronomy Observatories, USA.

be enormously greater: in fact the neutrino–baryon number ratio would be comparable to N_γ/N_B.

Particle physicists do not rule out massive neutrinos. In fact, according to some theories, neutrinos may exist in two states, light and heavy. The light neutrino may have a mass energy of a few electron volts, while its heavier counterpart may be several times as massive as the proton. The latter component of the doublet would be an awkward relic, much like the monopole. So we must assume that it has a lifetime short compared to the present age of the universe. The lighter component is expected to survive, on the other hand.

Coupled with the fact that neutrinos may exist in at least three flavours, ν_e, ν_μ, and ν_τ, and that the masses of ν_μ and ν_τ may be higher than the mass of ν_e, however, even the lighter relic neutrinos may turn out to be embarrassing for standard big bang cosmology. For example, if all three species survive to this day, then their combined mass has only to exceed the energy equivalent of 100 eV for them to have sufficient density to 'close' the universe. (The Hubble constant used here is with $h_0 = 1$. For $h_0 = 1/2$, this figure is reduced to a quarter of its value.) In this estimate, antineutrinos are also included, so they also contribute equally to the above limit.

Let us see why too many surviving flavours of massive neutrinos are embarrassing for the big bang. Recall from Chapter 2 that the age of the universe depends on two parameters, h_0 and Ω_0. For three species of massive neutrinos, each with a mass energy of 30 eV at least, the total contribution to universal density takes Ω_0 to 1.8 for $h_0 = 1$. (See Fig. 5.6.) The age of such a universe would be only ~6 billion years, barely exceeding the age of the earth and the solar system, and falling far short of even the most conservative estimates of the age of our galaxy—~12 billion years. Indeed, according to some age estimates of stars in globular clusters, galactic ages may be as high as 17 billion years.

What is the experimental status of measurements of neutrino masses? To date there have been several attempts, and a positive result is claimed by the Russian group headed by V. A. Lyubimov in Moscow. After careful analysis of their

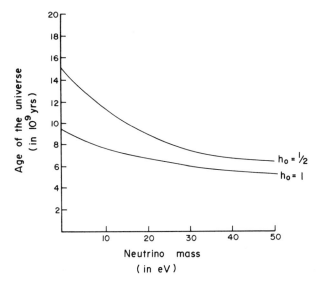

Fig. 5.6. The age of the universe plotted against the minimum mass of massive neutrinos, assuming that there are three neutrino flavours and that these dominate the dark matter in the universe.

experiment, this group claimed that the neutrino mass (for ν_e) is typically 30 eV, with the experimental uncertainty putting the range as ~16–40 eV. Clearly, a value in this range may be too high for the standard big bang model to be viable, especially if neutrinos of other flavours, with even greater masses, are also assumed to survive. In Chapter 6, when we review dark matter in general, we will come back to another problem associated with massive neutrinos.

The number of flavours of surviving neutrinos is also important. Referring back to the discussion in Chapter 3, we note that the primordial helium abundance is determined largely by the neutron–proton number ratio at the time when these two particles dropped out of thermodynamic equilibrium. If the number of neutrino flavours contributing to the expansion of the universe is increased—that is, if the g-factor, which includes all the effective spin states of participating particle species, is increased—then so is the rate of

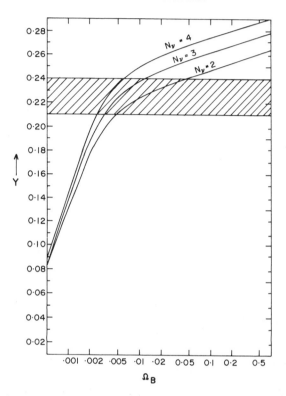

Fig. 5.7. The fraction of primordial helium, Y, goes up with the number of neutrino flavours. The shaded region marks the observed range of Y. Ω_B here stands for the baryonic density parameter, and Hubble's constant is taken as 100 km s^{-1} Mpc^{-1}.

expansion. Now thermodynamic equilibrium between neutrons and protons lasts until their rate of mutual interaction remains at a suitably high value compared to the universal expansion rate. Therefore, with more neutrino favours, thermodynamic equilibrium is lost sooner—that is, at a higher temperature. A look at the formula giving the ratio of N_n to N_p on page 91 tells us that if the equilibrium is lost at a higher temperature, then this ratio is higher, and so there is more helium in the universe.

Fig. 5.7 shows the observational situation *vis-à-vis* theory.

Already there are indications that with three neutrino flavours, the mass fraction, $Y = 0.25$, of primordial helium in a big bang nucleosynthesis is too high relative to observations. It would appear, therefore, that if the hot big bang picture is right, it severely limits the number of neutrino flavours—at any rate at the time of nucleosynthesis. This restriction is an example of how cosmology may limit the parameters of a particle theory.

The problem of galaxy formation

It is well known and accepted that the Friedmann models are idealizations of what the universe is like on the large scale. The homogeneity and isotropy of these models tell us that today on the scale of 10^{26} cm upwards, say, these properties apply to the universe. On smaller scales, however, there are galaxies, clusters of galaxies, and even larger superclusters of galaxies. How did these inhomogeneities come about?

Evidently these are also relics of some early epoch, when small fluctuations from the smooth Friedmann picture came about, fluctuations which grew to galaxies, clusters of galaxies, and so on. Many detailed studies of this process have turned into blind alleys. In 1900, Sir James Jeans introduced the notion of a characteristic minimum mass of a typical inhomo-geneity which must be present in a gaseous medium if it is to grow. In Jeans's argument, a mass larger than this critical mass commands a gravitational contracting force strong enough to withstand the disrupting thermal pressures in the medium.

The extension of these ideas to the expanding universe during the recombination epoch (see Chapter 3) leads to a critical Jeans mass on the order of a million solar masses. The magic number of a hundred billion solar masses (the mass of a typical galaxy) does not arise in the calculation in any natural fashion. Detailed studies in 1946 by E. Lifshitz showed that there was no satisfactory way that perturbations in the expanding Friedmann models could grow to galactic size.

The discovery of the microwave background and the growing realization through numerous observations that it is homogeneous on very small scales brings its own problems. For example, if we express by $\delta\rho$ the small inhomogeneity in the cosmic average matter density, ρ, then in a simple adiabatic cooling process—cooling by expansion, with no heat loss from the system—the ratio $\delta\rho/\rho$ is three times the temperature fluctuation ratio, $\delta T/T$, in the background. So far observations have succeeded only in placing upper limits on $\delta T/T$, of less than 10^{-4}. Thus the density fluctuations are too small to be capable of growing into galaxies—unless the picture of adiabatic cooling is wrong altogether.

Many astrophysicists believe that the above picture based on the Jeans mass may be too naïve, and that the very early universe may provide the clue to the problem. A result which appears to be borne out by empirical studies of the currently observed inhomogeneities is that they are scale-independent —that is, the ratio $\delta\rho/\rho$ does not seem to single out any linear scale, such as that of the size of a typical galaxy or cluster. Such a 'scale-invariant spectrum' was considered independently in the early 1970s by E. Harrison and Ya. B. Zeldovich in their models of galaxy formation. Could such a spectrum have arisen in a natural way in the very early universe?

The flatness problem

From the physical problems we now turn to a geometrical one, first pointed out in 1979 by R. H. Dicke and P. J. E. Peebles, and later highlighted by A. Guth in 1980. To appreciate it, let us go back to the first of the two equations on page 59, which we rewrite here as

$$\frac{\dot{S}^2}{S^2} + \frac{kc^2}{S^2} = \frac{8\pi G\rho}{3},$$

where S is the expansion factor, $\dot{S}$ its rate of change with time, ρ the density of matter, and k the curvature parameter which takes the values 1, 0, or -1. Suppose first that $k = 0$, so that we are in the flat space which expands with time. Write H for the Hubble constant and Ω for the density parameter; then

$$\Omega = \frac{8\pi G\rho}{3H^2}.$$

In the 'flat' case we of course always have $\Omega = 1$. If, however, $k \neq 0$, the above dynamical equation becomes

$$\Omega = 1 + \frac{kc^2}{H^2S^2} = 1 + \frac{kc^2}{\dot{S}^2}.$$

Notice that as we approach closer and closer to the big bang epoch, $\dot{S}$ increases, and the difference between Ω and 1 narrows down to zero. The present-day uncertainty in the value of Ω may be expressed by the inequality $0.1 \lesssim \Omega_0 \lesssim 5$. This range of values encompasses all three versions $k = 1, 0, -1$ of the Friedmann models, and appears very broad in the present context. Yet all the models falling within this range had Ω-values in narrower ranges in the past. If one takes the view that the nature of the universe was determined at the GUTs epoch, then the allowable values of Ω at that epoch were bunched close to the value $\Omega = 1$ with a spread as small as 10^{-50}. In other words, the universe was somehow very finely tuned close to $\Omega = 1$ at that epoch. If it were not so finely tuned—if, say, the spread $\Delta\Omega$ close to $\Omega = 1$ were on the order of unity at the GUTs epoch—then for $\Omega > 1$, the universe would have collapsed to a singularity, or for $\Omega < 1$, dispersed to infinity in a time of the order of 10^{-37} s. Thus fine tuning very close to $\Omega = 1$ was essential for the survival of the universe to the present state.

The fine-tuning problem, illustrated in Fig. 5.8, may be understood by means of an analogy: the difficulty of shooting at a small target from afar. The permitted range of deviation from the true direction becomes narrower and narrower as the marksman moves farther and farther away from the target. Likewise, the earlier we place the epoch when the curvature of the expanding space in the universe became fixed, the more finely tuned Ω had to be close to the 'flat' value of 1. In Chapter 6 we shall refer to the 'Planck epoch', up to which the universe may have been subject to quantum gravity. This quantum era would have lasted for a brief duration of $\sim 5 \times 10^{-44}$ s after the big bang. If the state of the universe were determined then, rather than at the later GUTs epoch, then

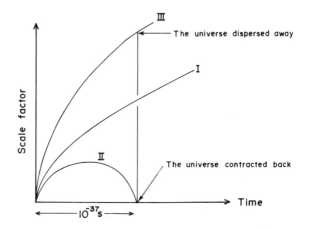

Fig. 5.8. If the geometry of the universe were decided at the GUTs epoch, the characteristic time for the Friedmann models should be of the order of 10^{-37} s. That is, models of classes II and III should respectively contract back to singularity or expand to infinity over this time-scale. The exception is the class I model (the Einstein–de Sitter model) which does not have a characteristic time-scale.

the required fine tuning would have to be even more precise (to 1 part in 10^{55}).

One may dismiss the problem by saying that there was fine tuning because the universe happened to start that way. But, as with the horizon problem, this resolution, by invoking highly special initial conditions, is unsatisfactory from the physicists' point of view. Is there a *physical* way of understanding this important aspect of the universe?

Thus we have an impressive and formidable list of problems which confront the big bang cosmologist who dares to apply the laws of physics to the study of the very early universe. In the rest of this chapter we will describe a scenario which modifies the Friedmann models considerably in this very early phase, in such a way that some, at least, of these problems disappear.

The Inflationary Universe

The word 'inflation', so painfully familiar to the common man attempting to make ends meet, was introduced into cosmology in 1980 by Alan Guth. The 'inflationary universe', or rather, the inflationary phase in the very early universe, owes its origin to the curious effects of phase transition when a spontaneous breakdown of symmetry occurs.

In the preceding chapter we discussed at length the phenomenon of spontaneous breakdown of symmetry and its role in theoretical attempts to understand the unification of physical interactions. Before we proceed to outline the inflationary scenario, let us recapitulate its salient features.

Recall first the apparently strange behaviour of a bar of ferromagnetic material at the Curie temperature. As the temperature of the bar is lowered through this value, there is a change in the available states of least energy. From a state which was manifestly isotropic, the bar finds that two states of even lower energy become available to it below the Curie temperature, and it takes up one of them, thereby apparently destroying the initial symmetry with respect to all directions.

Next recall the somewhat analogous phenomenon in GUTs. For example, in the $SU(5)$ theory, a phase transition takes place below a critical temperature of $\sim 10^{15}$ GeV. That is, below this temperature the $SU(5)$ symmetry between all component interactions apparently breaks down, and $SU(5) \rightarrow SU(3) \times SU(2)_L \times U(1)$. The right-hand side indicates the three component interactions as we know and study them at low energies: the strong, the weak, and the electromagnetic interactions. In the gauge-theoretic picture of Chapter 4, this change-over is associated with Higgs bosons, which mediate the interactions. For these bosons, the nature of the 'state of lowest energy' changes below the critical temperature of $\sim 10^{15}$ GeV, and lower energy states become available, which are different from the one to which they were accustomed before. In the jargon of quantum field theory, the 'true vacuum' shifts to lower energy at this change-over.

As described in the context of the Weinberg–Salam theory

of Chapter 4, this state of true vacuum below the critical energy apparently lacks the symmetry of the earlier vacuum wherein all Higgs boson fields had zero values. In this new vacuum state some of the fields have non-zero values. This situation is characteristic of gauge theories in general, and it is also valid for GUTs at the critical energy of $\sim 10^{15}$ GeV.

Let us now examine this particle-theoretic picture against the background of the very early universe. Recall the time–temperature relationship on page 151. Since we have expressed temperatures in GeV units, it is easy to read off the time corresponding to an energy of $\sim 10^{15}$ GeV. For $g = 100$, we get $t \sim 10^{-37}$ s. At this epoch, as the universe expanded further, the average particle energy would drop below this critical GUT energy, and a spontaneous breakdown of symmetry would be expected to occur. This is where Guth introduced the novel scenario of inflation.

The cooling of the cosmic material and the change in its physical behaviour at this critical juncture may be compared with the cooling of steam and its condensation to water, or with the cooling of water to ice. In the technical jargon of the condensed-matter physicist, these changes of state are called 'phase transitions', a term which describes very well what was happening to material at this juncture in the very early universe. The important aspect of phase transition invoked by Guth is called 'supercooling'.

The steam–water–ice example tells us what supercooling is. For example, steam can be supercooled below the boiling-point of water (normally 100°C) so that even at this lower temperature it retains its gaseous state. Likewise water can be supercooled and kept in liquid state well below its freezing-point. The supercooled state is unstable, however, since the physical system has a higher energy in this state than in the normal state, and nature allows (and encourages) transition to the state of lower energy. So with a slight disturbance the supercooled steam condenses to bubbles of water, and the supercooled water freezes to chunks of ice.

Whether (and if so, how) cosmic material might behave similarly and nucleate into bubbles depends on the physics of GUTs. Supercooling corresponds here to the material retaining its symmetric state (with all Higgs boson fields zero), even *below* the critical energy of 10^{15} GeV. This state is appropriately called a 'false vacuum', the true vacuum of lower energy being one of broken symmetry.

We may liken the two vacua to upper and lower valleys in a mountain terrain, the upper valley surrounded by mountains being the false vacuum. For an inhabitant of this valley the (false) impression is created that he is located at the bottom of the mountains. Unless he climbs over the surrounding peaks and looks beyond, he cannot know that a much lower terrain exists. If he does not have the energy resources to make this climb, he is for ever confined to his mountain Shangri-La.

That is what would have happened to the supercooled matter in the very early universe if its behaviour had been regulated by classical Newtonian physics. We may visualize the situation by seeing the energy of the system as playing the role of height in the mountain analogy. The bulk of the energy at this stage is carried by the Higgs bosons. The theory of the Higgs fields studied by Guth near the false vacuum showed that tall potential barriers surrounded it, and that the true vacuum of lower energy lay beyond these barriers. Unless particle energies exceeded the energies associated with these barriers, classical physics would not allow a change-over to the true vacuum state. However, quantum physics, which supersedes classical physics at such small scales and high energies, does allow the transition to take place.

It takes place by means of 'tunnelling', a term which can be understood using our mountain analogy. For if one cannot reach the other side of a tall mountain by climbing over it, technology may permit one to tunnel through it and reach the other side all the same! The rules of quantum mechanics allow motion across potential barriers which would be prohibited by classical Newtonian laws of motion. (See Fig. 5.9.) Of course, even in quantum mechanics, barrier penetration, though possible, is somewhat rare. Detailed calculations are needed to compute the probability that a particular barrier is penetrated. The taller the barrier, the lower the probability of penetration.

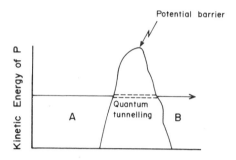

Fig. 5.9. Tunnelling in quantum mechanics allows a particle, P, to make a transition from state A to state B. In classical mechanics the intervening barrier forbids such a transition. Such tunnelling allows change-over from false to true vacuum.

Nevertheless, in our example, we have to allow for an event to happen somewhere, even if its probability is small. We therefore envisage small regions of space where tunnelling occurs, and explore the consequences.

At the completion of the tunnelling this small region finds itself in a state of true vacuum with lower energy than the surrounding false vacuum. Since there is now a dissimilarity in conditions between this region and that outside, the former has to readjust itself with respect to its surroundings. How does such a region behave?

As we have seen, the energies of the two vacua are not equal, the true vacuum having less energy than the false one. What about the pressures? Calculation shows that the false vacuum has negative pressure, while the true vacuum has zero pressure. For the time being we will take this strange property of the false vacuum as it stands and reserve comment for later. What matters most for our present discussion is that the pressure inside the region of true vacuum is *higher* than that prevailing outside in the false vacuum. The region therefore blows up like a balloon inflated by an air pump.

Guth compared this blowing-up process with that of a bubble of gas, and the bubble concept has stuck. The rates at which bubbles would form and expand are determined by the parameters of the particle theory used. The rate of formation is low, but the rate of expansion turns out to be extremely high, doubling its diameter every 10^{-34} s. The expansion of the linear scale of the bubble is thus exponential: $S = \exp at$, where a^{-1} is of the order of 10^{-34} s. This expansion, termed 'inflation', would last until the transition to the broken symmetry phase was complete.

When exactly is the phase transition complete? In Guth's original scenario, this state was thought to have been reached when every point in space was almost certainly inside a typical bubble. To assess when this would happen, one must compute how the probability of a typical point in space lying outside a bubble changes with time. Taking into account the above two rates, one finds that this probability, p, decreases exponentially with time:

$$p(t) \propto \exp\left(-\frac{t}{\tau}\right),$$

where τ is of the order of 10^{-32} s. This is the duration of the inflationary phase, the interval of time over which the scale factor S increased by the fantastic factor of $\sim 10^{50}$.

Let us now explore further the *cause* of this exponential expansion of the supercooled material of the universe. The Einstein equations of general relativity, which we mentioned in Chapters 2 and 3, take the following form in the present context:

$$\frac{\dot{S}^2 + kc^2}{S^2} = \frac{8\pi G}{3c^2}(u_r + u_v)$$

and

$$2\frac{\ddot{S}}{S} + \frac{\dot{S}^2 + kc^2}{S^2} = -\frac{8\pi G}{c^2}(p_r + p_v),$$

where u_r and p_r are the energy density and pressure of the relativistically moving particles in the universe, with $u_r = 3p_r > 0$, and the new terms u_v and p_v are the energy density and pressure of the false vacuum, with $p_v = -u_v < 0$.

Now the quantities u_r and p_r tend to decrease with the expansion of the universe as $1/S^4$, while p_v and u_v stay constant as long as the false vacuum is maintained. This can be seen from the above equations. After a little mathematical manipulation we get

$$\frac{d}{dS} (u_v S^3) + 3p_v S^2 = 0.$$

If the relation $u_v = -p_v$ is substituted in this equation, we get $u_v =$ a constant. Naturally u_v and p_v tend to dominate the behaviour of the solution of the equations. Further, these terms also dominate the curvature term, k/S^2 which (in the cases $k = \pm 1$) decreases as S increases. Writing

$$\frac{8\pi G}{3c^2} u_v = a^2,$$

the ultimate form of the two field equations is:

$$\frac{\dot{S}^2}{S^2} = a^2 \quad \text{and} \quad 2\frac{\ddot{S}}{S} + \frac{\dot{S}^2}{S^2} = 3a^2,$$

and it is easy to verify that S has the exponential behaviour described earlier.

The reader may notice here an echo of the classic cosmological solution obtained by de Sitter back in 1917 and described in Chapter 2. The 'motion without matter' of de Sitter is replaced here by the notion of the false vacuum, with the λ-term effectively arising from the energy density of the false vacuum. Of course, what we are highlighting here is the similarity of the functional forms of the equations and their solution in the two cases; the actual magnitudes of the λ-term and the corresponding term in the above formulation are quite different. The term in the inflationary scenario is about 10^{108} times as large as the λ-term in the cosmological setting imagined by de Sitter and Einstein.

In the final chapter we will discover an even more striking similarity between the inflationary scenario and another important cosmological theory introduced in 1948—namely, the steady state model of Hermann Bondi, Tommy Gold, and Fred Hoyle.

To complete our description of Guth's inflationary scenario we now need to consider one more physical consequence. Pursuing our analogy of supercooled steam, notice that when the steam condenses to water, it releases its so-called latent heat—that is, the heat required to convert water to steam by boiling it. In the phase transition from false to true vacuum, the energy difference between the two states is likewise released. This extra energy reheats the matter in the bubble, creating a large number of fast-moving particles. Guth's expectation was that the reheating would take the temperature in the bubble to about 10^{14} GeV.

This inflated region made up of bubbles now forms the starting-point for expansion as in the Friedmann model, for the vacuum energy effect has now disappeared. Thus the inflationary phase was a transient phase during which the expansion of the universe departed from its Friedmann-type behaviour ($S \propto t^{1/2}$). The scenario is illustrated in Fig. 5.10. Although the universe returned at the end of this phase to the Friedmann-like form, inflation had produced non-trivial changes in its composition. We will next see how far these changes allow us to solve the outstanding problems of the very early universe.

The Successes of the Inflationary Scenario

Guth's inflationary scenario for the very early universe was seen to possess several advantages over the standard Friedmann picture. To begin with, the horizon problem of the very early universe is solved by inflation. Recall that the present large-scale homogeneity of the observable universe, when traced back to the GUTs epoch, turned out to embrace an area some 10^{26} times larger than that compatible with the horizon size of that time. It was therefore hard to understand how regions well beyond one another's range of communication came to have exactly similar physical properties.

If we introduce inflation into this picture, the situation

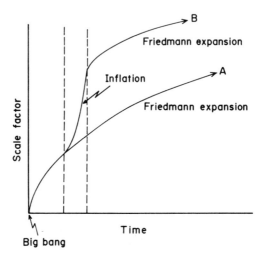

Fig. 5.10. The lower curve, A, represents the standard Friedmann expansion. The upper curve, B, represents exponential expansion (inflation) during the period indicated by dashed lines, followed by Friedmann expansion. (The curves are not drawn to scale, nor are the numbers given anywhere. These depend on the details of how the phase transition occurs.)

changes radically. The horizon size immediately prior to inflation gets blown up by a factor of 10^{50}. So the homogeneity which one would expect to hold within the pre-inflation horizon, would now extend to a region 10^{50} times larger after the inflation was over. This factor exceeds by a large margin the shortfall of 10^{26} encountered before. What is more, we can now assert that not only would we expect the observable universe today to be homogeneous on the large scale, but we would also expect to see it as a tiny speck in a homogeneous universe of considerably larger size.

This blowing-up of the horizon also resolves the magnetic problem, provided we can ensure that monopoles were formed during or prior to inflation, not after it. For our earlier computation of the monopole number density has to be corrected—that is, reduced—by the factor through which the volume of the bubble was increased by inflation *after* the monopoles were formed. The domain wall problem is similarly eliminated, since the walls are now blown far away from a typical point in the bubble.

The most attractive feature of the inflationary model is its resolution of the flatness problem, however. Recall that the flatness problem arose from the comparison of the dynamic term $\dot{S}^2/S^2$ with the curvature term kc^2/S^2 in Einstein's equations, and from the realization 'that at the GUTs epoch the latter had to be as small as 1 part in 10^{50} of the former if the corresponding ratio for the present universe was to fall within the current observational range.

If we take the inflationary scenario seriously, then we find that with $S \propto \exp at$, the dynamic term stays constant, while the curvature term declines exponentially. Indeed, the decline of the curvature term is so rapid as to shoot past the fine-tuning requirement of 10^{-50}. That is, immediately after the inflation is over, the subsequent Friedmann expansion would take over with a curvature term which is totally negligible. The clear prediction which emerges, therefore, is that the present state of the universe should be indistinguishable from the 'flat' $k = 0$ model; so the present value of the density parameter should be very close to unity.

Later Inflationary Models

Even at the time when he proposed the inflationary model, Guth was aware of some of its possible shortcomings. Later investigations were to accentuate these shortcomings, creating problems which seemed insurmountable. Yet the central attractive feature of supercooling during the phase transition and inflation due to the negative pressures of the false vacuum prompted new versions of the inflationary scenario, rather than its abandonment. By the end of 1981, A. D. Linde in Moscow had proposed a new approach which came to be known as the 'new inflationary model'. The same ideas were put forward independently at around the same time by A. Albrecht and P. J. Steinhardt in the United States. Before describing this new version, let us briefly see why the old version ran into difficulties.

The crucial flaw in Guth's version turned out to lie in the estimate of how close to one another the various bubbles created by phase transition would come. The expectation was that eventually the bubbles which formed randomly under phase transition would collide with one another. But this expectation was not realized: the bubbles tended to cluster in groups around the largest in a group, with each group lying well away from its neighbours.

Why was this a problem? Because theory showed that the energy released in phase transition would tend to reside on the surfaces of the bubbles, with the largest energy being on the walls of the largest bubble of the typical group. For 'reheating' to take place it was essential for the bubble walls to collide. Collisions occurring frequently would release and redistribute the energy residing in individual units, and without such a redistribution, reheating of the inflated universe could not be achieved. And without such a democratization of the energy structure, the universe today would be extremely inhomogeneous and anisotropic.

Even the new inflationary model was not altogether successful. In its original version it adopted a different type of energy potential linking the false and the true vacuum, a potential function which had been studied in particle physics by S. Coleman and E. Weinberg back in 1973. To understand why this new potential function was considered advantageous over the one chosen by Guth, let us go back to the mountain–valley analogy discussed earlier.

In Guth's scenario the false vacuum lay in a high valley surrounded by a barrier of mountains which descended to the much lower level of the true vacuum on the other side. So quantum tunnelling was needed to penetrate the barrier. Once tunnelling took place, bubbles formed. The growth of a typical bubble after formation was not rapid enough to guarantee collisions between bubbles, however. In the new scenario the false vacuum lies on a tall plateau which slopes down extremely gently for a while, then descends very rapidly down

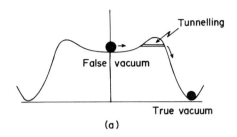

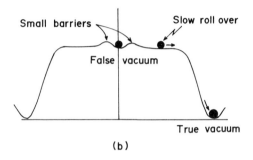

Fig. 5.11. The potential function for the original inflationary universe (a) and a further modification thereof which includes a shallower valley at the top, followed by a slow roll along a plateau (b).

a canyon to where the true vacuum lies. Imagine a ball travelling down such a track. It would begin by rolling very slowly down the side of the plateau, but once it reached the canyon, it would descend very rapidly. Hence most of the time during which it is in motion is spent in a slow roll from the highest level. Fig. 5.11 illustrates this scenario.

This behaviour translates into the physics of phase transition thus. While the gradual roll at the high level is going on, conditions similar to those of a false vacuum operate within the region. Thus the region grows exponentially during time, doubling its size every 10^{-34} s or so. While this is happening, the phase transition proceeds very slowly, with the Higgs fields growing very little from their initial zero value. When the rapid descent to the true vacuum begins, the Higgs fields grow fast and attain their final non-zero values. This is when inflation would stop, and Friedmann expansion would take over. The slowness of phase transition now permits inflation to go on for a much longer time than in the earlier version.

Thus, instead of several isolated groups of bubbles nucleating, as in Guth's scenario, we now have a single large region which today would vastly exceed the size of the observable universe.

How would the reheating take place in the new scenario? We do not have to resort to collisions in this picture. The energy release occurs through rapid oscillations and subsequent decays of the Higgs fields/particles. Being unstable, the heavy Higgs bosons would be converted into lighter particles, which would be produced in profusion. It is these lighter species which would achieve thermodynamic equilibrium through rapid collisions.

This new inflationary model underwent further metamorphosis with another theoretical change in the potential function. In this case (illustrated in Fig. 5.11) there was a small mountain barrier around the high valley representing the false vacuum, followed by a plateau, as in the earlier version of the new model. Thus modest tunnelling was required before the slow roll could begin. This picture thus combines certain features of both Guth's model and the new model. Thus bubbles may form and grow exponentially. The fraction of space filled with bubbles may approach unity as time passes, but all the same, the gaps between bubbles (representing the false vacuum state) will also grow in volume. In these gaps new bubbles would appear.

The new inflationary model thus solves the reheating problem which bugged the original model, and also resolves the horizon and flatness problems. Like its original counterpart, it solves the domain wall and magnetic monopole problems, too. Somewhat unexpectedly, it also does well in explaining the spectrum of inhomogeneities which we mentioned in the context of galaxy formation.

This last feature is related to inherent inhomogeneities of the Higgs field prior to inflation. As is to be expected, quantum theory does not allow a completely smooth Higgs field, even in the vacuum state of zero value. The 'zero' is an average, and there are bound to be fluctuations. These inhomogeneities are of course on a very small scale initially (that is, prior to inflation), but they grow in size by inflation. They are transferred from the Higgs field to matter after the Higgs bosons decay. Calculations show that the inhomogeneities in matter distribution are scale-independent, much as in the phenomenological models of Harrison and Zeldovich.

However, at this point of relative success, problems begin for the inflationary scenarios! The actual magnitudes of inhomogeneities are far too large compared with observations. For example, as we noted earlier, the extraordinary smoothness of the microwave background suggests that the density fluctuations, $\delta\rho$, compared to the average density, ρ, cannot have been more than 10^{-3} (even at a very conservative estimate). The new inflationary scenarios described above produce $\delta\rho/\rho$ values ten to a hundred thousand times too high. Thus on this count alone, even the new inflationary scenario based on well-understood GUTs seems doomed to failure.

Moreover, the purist may raise two further objections on aesthetic grounds. First, the slow roll so crucial for the new inflationary scenario to work at all is based on a very special choice of parameters for the potential function. This means that unless the parameters of GUTs are fine-tuned to these values, the model will not work. Thus it seems that, in order to get rid of the fine tuning implied by the flatness problem in the standard model, we have been led to a scenario which requires its own fine tuning.

The fine tuning appears in the second objection also. Recall that on the basis of cosmological observations the λ-term is believed to be very small today, smaller by a factor of 10^{108} than the λ-term governing inflation. The inflationary λ-term arises from particle physics, and is supposed to disappear after phase transition. We therefore require the disappearance of the inflationary λ-term to be exact to 1 part in 10^{108}. If this were not the case, we should see its effects today in the form of a dominating cosmological λ-term.

The inflationary scenarios evade, rather than tackle, the problem of the space–time singularity associated with the big bang. In the models so far described, the singular epoch occurs *prior* to inflation and cannot be dealt with by it. The best that inflation can do is to 'erase' any memory of the past history of the universe by exponential expansion; this is to argue that, no matter what the state of the universe before inflation began, what we now see in the observable part of the universe will be insensitive to those conditions.

This point of view has led to the notion of inflation out of 'chaos'. Imagine the universe in a highly non-uniform state with some parts expanding, some contracting, some hot, some cold. The hot expanding regions will cool, and as their temperature drops through the GUT phase-transition value, inflation will begin inside them. The large-volume expansion by inflation will eventually take in the chaotic regions which did not inflate. So for the miniscule space which expanded to our observable part of the universe, the chaotic conditions hardly matter.

Nevertheless, if the chaotic pre-inflationary universe was subject to the Einstein equations of gravity, it must have had a singular past. The singularity problem remains in the absence of new inputs into classical gravitational theory. We will discuss such an input in the next chapter.

It appears therefore that the problems of the very early universe have not all gone away with the advent of the inflationary scenarios. Nevertheless, the successes of the inflationary models prompt some cosmologists to persist with the basic notion, but with new and more esoteric particle theories to guide them. There are others who believe that the inflationary scenario itself must be abandoned in favour of some new concept. In the chapter which follows we will highlight some of these more adventurous approaches to the very early universe.

6

Further Adventures on the Physics–Cosmology Frontier

> It is a capital mistake to theorise before one has data.
> Insensibly one begins to twist facts to suit theories
> instead of theories to suit facts.
>
> Sherlock Holmes

The inflationary scenario described in Chapter 5 was the offspring of a marriage between GUTs and classical Friedmann cosmology. The difficulties encountered by the scenario are a reflection of the fact that the marriage has been a difficult one, if not entirely unsuccessful. At the same time, the successes of the inflationary universe prompt one to proceed further along the road of exploring the *very* high energy physics in the *very* early universe. Indeed, since the early 1980s, ideas more adventurous than those embodied in GUTs have begun to gain currency among theoretical physicists. Some of these ideas are finding echoes in studies of the very early universe. The inflationary scenarios may also acquire further variants with inputs from the new ideas.

At the time of writing, no broad consensus—let alone a single line of attack—has emerged from these thought-adventures. Theoretical physicists are fond of 'thought-experiments', which explore by imaginary experiments the physical behaviour of a system under conditions hard to realize in practice. By the term 'thought adventure', we push this concept even further along the road to intuitive incredibility. Our purpose here, therefore, is not to present a coherent picture of the very early universe—since none exists—but to convey to the reader the excitement of work on this frontier area where physicists have tended to shed the inhibitions which would normally constrain their approaches in more routine investigations.

Supersymmetry and Supergravity

Recall from Chapter 1 Aristotle's penchant for circles. His insistence on circular motions as those preferred by nature reflected a belief shared by philosophers of his time (and even earlier times) that natural processes are governed by simple underlying principles. The Greek approach was to 'discover' those principles through intellectual exercise, and then to see how their effects might translate into observable data. The epicyclic theory started accordingly with a firm conviction of the reality of Aristotle's circular trajectories, but subsequently had to resort to increasing complexity of description in order to sustain that picture. Its subsequent overthrow by the discovery of Kepler's laws marked the starting-point of a scientific revolution wherein observations of natural facts or laboratory experiments became the ultimate deciding factors *vis-à-vis* physical theory. This mode of operation has paid rich dividends in providing us with the science we have today.

But in high energy physics, or rather in *very* high energy physics, this *modus operandi* has shown its limitations. As we discovered in the previous two chapters, there are *no* laboratory methods capable of energizing particles to the levels needed to test the various GUTs. This limitation has forced theoreticians of today to go back, somewhat reluctantly, to the methods of their ancient Greek counterparts, to try to discover the underlying principles of nature from criteria of symmetry and beauty.

Supersymmetry and its natural offshoot, supergravity, are modern theories which would have delighted Aristotle & Co. Like the adjective 'grand' in GUTs, the prefix 'super' is partly a reflection of the circumstance that theoreticians today live in the media-oriented world of the superstar, the superman, and the supercomputer. More seriously, it indicates a belief that we are talking about a symmetry which transcends all other symmetries in particle physics.

In a nutshell, supersymmetry is the symmetry between fermions and bosons. To elaborate this statement further, we recall from Chapter 4 that these two types of particles are distinguished by their statistical properties, which in turn are related to their intrinsic spin. Bosons are particles with integral spins, 0, 1, 2, and so on, while fermions have half-odd-integral spins, 1/2, 3/2, and so forth. We have also found that our description of particle physics so far has consisted of fermionic particles interacting by exchange of bosonic ones. Thus the electromagnetic interaction is mediated by the photon (spin 1), gravitation by the graviton (spin 2), and the electroweak interaction, QCD, and GUTs by bosons of various kinds. The interacting particles—electrons, protons, neutrinos, quarks, and the rest—are all fermions.

Before supersymmetry came on the scene, particle theories maintained a rigid distinction between these two species of particles. A symmetry transformation which changed one particle into another, and the associated particle reaction, always preserved the fermionic or bosonic nature of the inter-acting particles. By contrast, a supersymmetry transformation associates a fermion with a boson, and vice versa.

Take gravitons and photons, for example. Under a super-symmetric transformation a graviton would go into a fermionic particle of spin 3/2, and a photon into a particle of spin 1/2. What are these particles? Since no such particles are currently known among the family of fermions, new ones have had to be invented: the 'gravitino' and the 'photino'. Like their bosonic counterparts, they are electrically neutral and are associated with the gravitational and the electromagnetic interactions respectively.

Correspondingly, new bosonic counterparts have had to be invented for known fermions. Thus the 'selectron' (spin 1) corresponds to the electron, the 'sneutrino' (spin 1) to the neutrino, and so on. By and large the prefix 's' indicates the supersymmetric partner of a known fermion, while the ending 'ino' denotes the supersymmetric fermionic counterpart of a known boson.

Thus supersymmetric theories appear to have doubled at one

stroke the number of species which particle theorists must deal with. Is this the right direction in which to be moving, when the main aim is to understand the plethora of subnuclear particles in terms of a compact unified picture? Defenders of supersymmetry counter this argument in the following way.

First, supersymmetry, often shortened to SUSY, sets its sights on a goal more ambitious than that of GUTs. It aims at unifying *all* physical interactions, those included in GUTs as well as gravity. The mathematical formalism of quantum field theory, however, poses one immediate problem. The graviton has spin 2, whereas the photon and the other bosons in GUTs have spin 1; and it is not possible in the conventional framework to envisage transformations which alter spins. SUSY theories permit this change to take place via the gravitino; thus:

$$\text{graviton} \rightarrow \text{gravitino} \rightarrow \text{photon}.$$

Further, by removing the distinction between fermions and bosons, SUSY completely unifies 'matter' and 'interaction', as no lower symmetry has been able to do.

A further dividend which *may* come from SUSY is resolution of the renormalization problem of quantum field theory. In Chapter 4 we saw this problem as arising from a series of divergent integrals describing measurable physical quantities like electronic charge and the electron mass. The technique of renormalization gives a way of separating out awkward infinities and arriving at meaningful finite answers. Yet, in spite of the successes of this technique, the purist cannot help wondering whether a new, correct description might exist wherein such infinities are avoided altogether.

SUSY holds out the hope of such a description. The hope springs from the fact that the infinities associated with fermions (for example, the self-energy and the effective mass of the electron) are of opposite sign to those associated with bosons (for example, the vacuum polarization and the effective strength of the electronic interaction). So by bringing new 'sfermions' and 'bosinos' into the picture, it may be possible to cancel out the opposing infinities entirely. Although there is no rigorous demonstration of such a general result in the SUSY literature yet, workers in the field find that SUSY theories are 'remarkably friendly' when handling quantum divergences.

This so-called friendly behaviour of SUSY comes in handy when dealing with the 'hierarchy problem' of GUTs. This problem is of a theoretical nature, and owes its origin to the circumstance that between the typical energies of the electro-weak theory ($\sim 10^2$ GeV) and of GUTs ($\sim 10^{15}$ GeV) lies a big energy gap of some thirteen orders of magnitude. When studying particle interactions by the technique of perturbation expansion (the technique most often used in particle physics, and sometimes the only one available), this large energy gap results in 'instabilities'. That is, answers obtained by the technique are not immune to slight changes in the physical description. Only a repeated fine tuning of GUTs parameters can make the answers at all meaningful. This situation is largely avoided by SUSY with the new particles at its disposal. These particles offer extra contributions to the various reactions which tend to cancel out the awkward effects that previously needed fine tuning of parameters for their elimination.

There are some very general theorems in supersymmetric theories which display its aesthetic merits. For example, the energy of a system under SUSY can have only positive values. (This is something not guaranteed in quantum field theories without SUSY, a feature which has always proved to be an irritant.) Further, any state with positive energy must have more than one particle in it. Thus we must have at least a pair of particles, one boson and one fermion with spins differing by $1/2$. All particles belonging to such a pair, or multiplet, must have the same mass.

SUSY theories differ among themselves in the number of supersymmetric partners allowed in each state. For example, how many partners (photinos) a photon can have depends on the basic assumptions of the theory. In the $N = 1$ SUSY theory, there would be only one photino; whereas an N-extended SUSY theory would allow N photinos. A general theorem tells us that the particles in an N-extended SUSY theory will have spins of at least $N/4$. Now it is known that gravity cannot couple consistently with particles of spin $5/2$ or more. It follows, therefore, that a SUSY theory of gravity —that is, supergravity—cannot have $N > 8$. Similarly, since spin-$3/2$ particles pose problems for renormalizable field theories (see Chapter 4 for a discussion of renormalization) in flat space–times, such theories should have N at most equal to 4.

However elegant such theorizing might be, recourse to hard facts reveals the basic weakness of SUSY; for as yet, there are no hard facts to support any of the SUSY ideas. A cynic might say that nature seems to be blissfully unaware of how it should behave. For example, none of the SUSY partners have yet been observed. Experiments have failed to reveal any pairs, or multiplets, of particles with differing spin but the same mass. To get out of this impasse, SUSY theorists have to argue that SUSY can be realized only as a spontaneously broken symmetry. Theoretical arguments show that this can happen provided the vacuum energy is not exactly zero. Only then will the vacuum state be asymmetric with respect to SUSY transformations. Such recourse to breakdown of symmetry, however, seems to rob SUSY of the elegance and simplicity with which it started.

The masses of SUSY partners are typically expected to lie in the TeV range (1 TeV = 10^3 GeV). Thus experiments using the next generation of high-energy accelerators may meaningfully search for SUSY effects. (The existing proton–antiproton collider facility at CERN has looked for SUSY-relevant masses up to 100 GeV, but without success.)

Thus SUSY theorists are coming to cosmology to explore the effects of their theories on the very early universe. Although a particle mass of 1 TeV would place the typical SUSY effect much later than the GUTs epoch (see the time–temperature relationship on p. 151), SUSY inputs to inflationary scenarios would have been in effect earlier. We mentioned in Chapter 6 the main difficulty of GUTs inflation —namely, that it leads to density fluctuations, $\delta\rho/\rho$, far higher, by a factor of 10^4 or more, than what are consistent with present-day observations. By choosing SUSY masses suitably, it is possible to arrive at an inflationary picture wherein $\delta\rho/\rho$ is as small as one wishes.

Although this result may be claimed as a success for SUSY inflation, we should not lose sight of the fact that this is nothing more than a parameter-fitting exercise. SUSY theory simply brings in more free parameters which can be, and have been, adjusted to eliminate the $\delta\rho/\rho$ problem. There is no predictive power in such a procedure. A more credible performance would require the SUSY theorist to say *why* the parametric values chosen are of special significance for the SUSY theory *per se*—that is, irrespective of the context of inflation.

The reader may also wonder whether the inflationary scenario itself is ever going to deliver the goods. Since its inception it has already gone through many changes and modifications, somewhat like the epicycles produced by the ancient Greeks to prop up their geocentric theory of planetary motion. Nevertheless, in the next section we will refer to one more version of inflation.

Another cosmological effect of SUSY will be the nature of stable relics from the SUSY epoch. For the lightest SUSY particle must, by definition, be absolutely stable, simply because it cannot go into anything else. What is that stable particle? Candidates are the photino, the sneutrino, the gravitino, and so on, with the photino as the leading contender perhaps. Whatever the relic, it will form part of the so-called dark matter which we shall review later in this chapter.

To sum up then, SUSY remains an idea on the drawingboard which awaits experimental verification; its applications to cosmology are still at the embryonic stage, and their observational consequences have been put forward only tentatively. Supporters of SUSY advise patience, for, as they say, 'sRome was not built in a day.'

The Kaluza–Klein Theory

Just as de Sitter's 1917 model found an echo in the inflationary universe, so another idea from the early days of general relativity has met with renewed interest in modern times. This idea, proposed independently by T. Kaluza in 1921 and O. Klein in 1926, arose in connection with attempts at a so-called unified field theory, which sought to combine gravitation with electromagnetism in a unified framework.

The Kaluza–Klein theory (K–K theory for short) introduced an extra spatial dimension into Einstein's description of gravitation as a manifestation of curved space–time. Thus the enlarged space–time of the K–K theory had one time dimension and *four* space dimensions.

What did the extra space dimension do? Evidently it had to be different from the other three space dimensions whose existence and effects are manifest in so many ways; the extra dimension had to do with electromagnetic effects. For, in keeping with Einstein's approach to a unified theory, physical interactions had to be described through geometry. The electromagnetic interaction had to be related to the geometrical inputs via the fifth dimension of the K–K theory.

The geometry was assumed to be periodic with respect to the extra space coordinate, and the magnitude of the period was related to the electric charge. The periodicity may be visualized by analogy with a right circular cylinder. Fig. 6.1 illustrates it. The cylinder is formed by wrapping a plane sheet around an axis. If L is the circumference of the cylinder, all points on the sheet lying on a straight line perpendicular to the axis and separated by L and its integral multiples fall on one another. Thus, if the geometry of the original sheet is periodic, with length L as the period, all its features are captured by going round the cylinder just once.

The extra dimension therefore differs from the other three in that spatial measurements along it are finite, whereas those in the three regular directions can stretch to infinity. This finiteness of spatial extent is often described by the word 'compactification': through the introduction of periodicity, the space in that direction has been made compact.

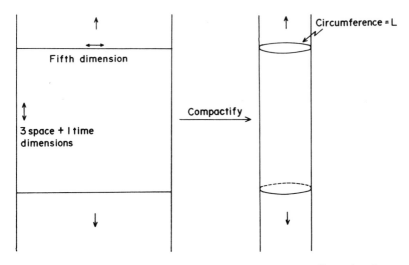

Fig. 6.1. The 'cylindrical' structure of the extra space dimension in the Kaluza–Klein theory.

The K–K theory had limited success as a unified field theory. However, its potentiality for introducing 'new physics' through extra dimensions has been appreciated in recent years in the renewed search for unification. In particular, the notion of compactifying the extra dimension has proved to be useful in describing the discrete parameters which characterize the internal degrees of freedom of particles.

Since the internal degrees of freedom, such as those described in Chapter 4, seem to follow symmetries described by group theory, the number and nature of the new dimensions are determined by group-theoretic considerations. Such considerations lead to K–K theories of eleven and even more dimensions!

For example, we may look upon the fifth dimension introduced by Kaluza and Klein for electromagnetism as corresponding to the abelian U(1) gauge group. To enlarge the scope of the original theory to include non-abelian symmetries, further new dimensions are added. The coordinates needed to describe the extra dimensions are then simply the parameters of the group in question.

This extended Kaluza–Klein approach has yet to prove its merits in terms of concrete verifiable predictions. Nevertheless, because of its intrinsic elegance, it is believed by many physicists to hold the clue to the as yet mysterious internal degrees of freedom of particles. Its dividend for cosmology is expected to be in terms of a modified (and hopefully workable!) inflationary scenario, which may be triggered by a spontaneous compactification of space along the extra N dimensions. Thus the universe is said to have originally had $3 + N$ non-compact spatial dimensions, all treated alike. At a critical temperature a breakdown of this symmetry is expected, with the compactification of N dimensions. In this process entropy would be released: at least, that is the hope if the scenario is to work!

Strings and Superstrings

The notion of strings comes into particle physics via the process of phase transition. Again, going by the analogy of the solidification of a liquid, we sometimes have elongated needle-like structures forming under this process, rather than spherical grains. In the GUT phase transition also, there is a strong possibility of forming strings instead of bubbles.

Strings are characterized by a fundamental length, L. In the 1960s, when strings were first invoked to describe hadrons, L was taken to be around 10^{-13} cm. More recently, the characteristic length of a string has been taken to be much smaller, on the order of the Planck length:

$$L_P = \sqrt{\left(\frac{G\hbar}{c^3}\right)} \cong 1.6 \times 10^{-33} \text{ cm}.$$

Because of the presence of G, we suspect that this string theory deals with gravity: indeed, it turns out that string theories contain a massless mode of spin 2, which may be identified with the graviton.

String theories can be quantized, provided the space–time has the appropriate number of dimensions. The first string theory to be quantized described only bosons, and it required twenty-six dimensions! It generated many strange results, and contained several conceptual problems, but it remains a useful exercise in understanding how the 'physics' of strings operates. A more popular class of theories use SUSY and require ten dimensions. Known as 'superstring theories', these are free of most of the problems which beset the earlier bosonic string theories.

In surveying these theories, it is easy to get lost in pedagogy and in the technical problems of 'tachyons' (particles which travel faster than light), 'ghosts' (negative energy modes, corresponding to nothing in the physical world), and 'loop diagrams' (which arise from perturbation theory). The cosmologist may ask, 'What is in it for me?' The answer is that strings and superstrings may be important in the very early universe, because such structures may act as seeds for forming galaxies. For example, A. Vilenkin and N. Turock have shown that the spectrum of inhomogeneities left by such strings is just the Harrison–Zeldovich spectrum mentioned in Chapter 5.

If strings remain as relics of the very early epochs, how much energy do they carry? The answer to this question depends on the particular string model and on the inflationary scenario used. For consistency, of course, we need the present density in the form of strings to be much less than the present matter density. An upper limit on the ratio of these two densities is obtained by the present upper limits on the fluctuations of temperature of the microwave background. This upper limit is about 10^{-3}, just large enough to leave another detectable effect, that of gravitational radiation: for strings will radiate gravitational waves. It is also likely that strings could have influenced the microwave background fluctuations at the recombination epoch, and have left detectable effects. It is too early to tell whether string theories will get a yes or no vote from cosmology.

Dark Matter and Structure Formation in the Universe

In Chapter 5 we mentioned briefly the problem of galaxy formation in the very early universe. A 'galaxy' represents a discrete structure, and the problem has been in arriving at a scenario whereby discrete units would emerge from a homogeneous background. Discrete structures exist in the universe at all levels, of course, from subatomic particles to giant superclusters. The cosmologist is concerned only with those in this hierarchy, which are at the galactic or higher levels. At all these levels he encounters the so-called dark matter, or missing mass, referred to in Chapter 5. It is now believed that problems of structure formation are intimately related to the nature of the dark matter which pervades those structures. Indeed, speculations on these subjects form a major part of current research on the early and very early universe.

Before we start speculating about these problems, let us outline some information obtained from astronomical observations and their reasonably conservative theoretical interpretation.

We have already noted that spiral galaxies show flat rotation curves, which imply (in terms of the inverse square law of gravitation) the existence of unseen gravitating matter. The dark matter is believed to exist in massive haloes around such galaxies. The data on galaxies of other types, of which elliptical galaxies form a major fraction, are consistent, if less definitive, with the conclusion that there are haloes around them too. Present observations indicate that the haloes effectively extend to about 3×10^5 light-years. In that case Ω_0, the present-day density parameter of the universe inclusive of dark matter, may be as high as 0.2. If the dark haloes should extend well beyond this distance and if the density should continue to taper off as the inverse square of the distance, then the haloes of neighbouring galaxies may well merge into one another, and the effective Ω_0 may approach the closure value $\Omega_0 = 1$.

The next step in this hierarchy is from galaxies to groups and clusters of galaxies. Data on random velocities of galaxies in a group or cluster imply the existence of non-luminous matter of density an order of magnitude higher than matter in luminous form.

As mentioned in Chapter 1, the next and largest scale of structure so far detected is that of the so-called superclusters.

The first of these was spotted by Gerard de Vaucouleurs in the late 1950s. This is the Local Supercluster, to which our Local Group of galaxies containing the Milky Way and the Virgo cluster of galaxies belong. The latter is the centre of the Local Supercluster, and is located about 15 h_0^{-1} Mpc away from us. It consists of a disc component of about 60 per cent of all luminous galaxies and a halo component of the remaining 40 per cent. A supercluster may be 50 h_0^{-1} Mpc in size, or even larger.

Advances in electronic technology have enabled astronomers to chart large-scale structure more and more effectively. Detection of faint galaxies and measurements of red-shifts have begun to produce a truly three-dimensional perspective on the universe. Thus, astronomers have learnt of 'filaments' and 'voids' which permeate the superclusters. Filaments are chains of galaxies which stand out against relatively sparse backgrounds. Voids are regions in which hardly any galaxies are visible. Such structures have characteristic dimensions of 30–60 h_0^{-1} Mpc. Fig. 6.2 shows filaments and voids in the Perseus and Pegasus supercluster.

Which came first, the superclusters or the galaxies? This is the chicken-and-egg problem which cosmologists are trying to solve. In the 'top-down' scenario the superclusters formed first, and the smaller structures fragmented out later; whereas in the 'bottom-up', or 'hierarchical clustering', scenario the galaxies were the first to form, at random places, and subsequently congregated in larger and larger clusters, leading to superclusters. The former view has been championed by Yakov Zeldovich, the latter by Jim Peebles.

The relevance of particle physics, GUTs and SUSY, inflation and the very early universe to the problem of structure began to be appreciated in the early 1980s. The basic idea behind all galaxy-forming scenarios has been the growth of initial density fluctuations, $\delta\rho/\rho$, until they exceed unity, when the 'non-linear' regime begins, with gravitational contraction of excessively dense regions. The process of growth and subsequent contraction is inhibited by the expansion of the universe. So the specific details of a scenario require

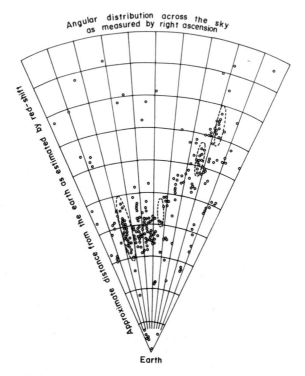

Fig. 6.2. Filaments and voids in the Perseus and Pegasus super-cluster. The circles represent galaxies, and those enclosed in dashed curves are clusters catalogued by George Abell. Figure adapted from that in Jack O. Burns, 'Very Large Structures in the Universe', *Scientific American*, July 1986, p. 40.

somewhat detailed knowledge of *what kind of matter* existed at *what earlier epochs*.

This is where confusion reigns. First, we have already ruled out baryonic dark matter (see p. 160). In Chapter 5 we also discussed massive neutrinos, which once constituted the only known alternative to baryons. Today, fortunately or unfortunately, the alternatives are far too many!

The jargon in this field talks of three types of dark matter: hot, warm, and cold. Hot dark matter consists of particles which were in thermodynamic equilibrium and moving with speeds close to that of light at the epoch when the last phase transition took place in the universe. This is believed to have happened at a temperature of about 100 MeV, when quarks finally disappeared into bound hadronic particles. Neutrinos belong to this class of dark matter, unless they happen to be too massive (with rest energies >100 MeV). The hot particles have number densities today of the same order as that of photons in the microwave background.

Warm dark matter consists of particles which interact more weakly with other matter than do neutrinos. So they decoupled from the other matter much earlier than neutrinos did. Their relic number densities are about ten times lower than those of hot dark-matter particles. Gravitinos are possible candidates for warm dark matter.

Before we come to the last class of dark matter, we will highlight one distinguishing property of all dark-matter particles which has implications for the kind of large-scale structures to which they might lead. The property is that of 'free streaming'—a measure of how far particles of a particular kind may travel freely during their time span of relativistic motion. If dark-matter particles travel a distance L, then their movements tend to wipe out inhomogeneities on scales *smaller* than L. For hot dark-matter particles L is around supercluster size. So one is led to the top-down version beginning with superclusters, because nothing smaller is allowed to grow. For warm dark matter the free-streaming criterion leads to structures at least as large as galactic haloes. Fig. 6.3 illustrates the three types of dark matter.

For cold dark matter there is practically no free streaming. Heavy stable neutrinos and axions are possible candidates for cold dark matter. Black holes of primordial origin, first proposed in 1975 by Stephen Hawking, are also possible dark-matter candidates, although not very popular ones. Another new species are 'quark nuggets'—that is, hypothetical chunks of quark matter containing roughly equal numbers of u, d, and s quarks. Such symmetric combinations are believed to be stable forms of matter, although none has yet been found in

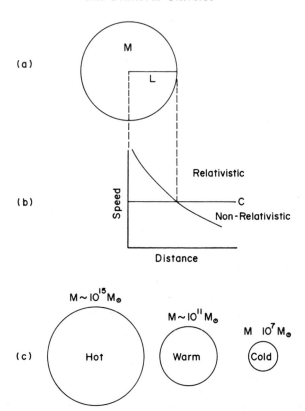

Fig. 6.3. *M* is the mass affected by dark matter free-streaming over a distance *L* (a), where *L* is the distance travelled by a dark-matter particle before it becomes non-relativistic (b). Typical values of *M* for hot, warm, and cold dark matter are shown in (c).

nature. Photinos can also be considered to be dark-matter candidates.

It is hard to do justice to the flurry of research activity on dark matter and structure formation in a general description like this, for various reasons. First, there is as yet no concrete result to talk about; second, the particle physics theories on which the arguments are based are themselves in a state of flux; and third, the astronomical view of the extragalactic

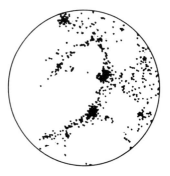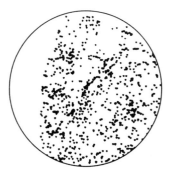

Fig. 6.4. The strong clustering that emerges from a typical massive neutrino scenario of galaxy formation (left) compared with the actual distribution of galaxies according to the Center for Astrophysics catalogue (right).

universe is itself improving year by year. The basic approach to the problem generally consists of the following steps. First, a scenario for the very early universe, with or without a particular brand of inflation, is devised; second, a particle theory (GUTs, SUSY, K–K, or something else) and the resulting recipe for dark matter—cold, warm, or hot, or a suitable mixture—is adopted; third, a prescription about how growth takes place is proposed, including thermodynamic constraints; and fourth, an analysis or simulation of N bodies is carried out by means of a fast computer.

The outcome of such an investigation is then matched with the observations of structure we outlined before. It must incorporate known features of the early universe, such as the primordial nucleosynthesis and the anisotropies of the microwave background. Certain scenarios may be ruled out as unworkable, and the hope is that the choice may eventually be narrowed down.

For example, the once popular massive neutrino (hot) dark-matter model has run into difficulties, in that the galaxies which are formed eventually in this scenario are too strongly clustered to be compatible with observations. The difficulty, illustrated in Fig. 6.4, can be circumvented, but it requires a great deal of fine tuning of the properties of neutrino species, their lifetimes and parameters. To quote Piet Hut and Simon White, who have worked with massive neutrino scenarios, 'There seems to be no natural reason why the parameters of real neutrinos should lie in the narrow ranges required to get a viable cosmology.'

Quantum Cosmology

The investigations of the very early universe described so far no doubt take us very close to the big bang epoch, $t = 0$. The GUTs epoch extending as far back as $t = 10^{-37}$ s already stretches the imagination well beyond the range of intuition. Nevertheless, it is possible to stretch our imaginations even further, by pushing the theoretical thought-adventure to an even earlier epoch, the so-called Planck epoch.

$$t_P = \sqrt{\frac{G\hbar}{c^5}} = 5.4 \times 10^{-44} \text{ s.}$$

Notice that three fundamental constants have gone into this formula. Although we have not yet said anything about the significance of this time-scale, the appearance of c, G, and $\hbar$ already indicates that it has something to do with the bringing together of quantum theory, represented by $\hbar$, and general relativity, represented by G and c. Indeed, this time-scale brings us face to face with what is considered to be one of the most difficult problems in theoretical physics. Like most questions which are difficult to answer, it can be stated simply: 'How is gravity quantized?'

Today physicists take it as axiomatic that however successful classical physics may be in describing natural phenomena, the ultimate description of nature has to be in terms of quantum physics. The three interactions electromagnetism, strong, and weak which GUTs seek to combine are already described in the quantum framework. Of these, the first started off as a classical field theory and achieved considerable success; yet the classical description proved inadequate in describing all observed electromagnetic phenomena. When the theory was quantized, it exhibited unsuspected richness, and it has successfully coped with experimental challenges.

First, let us ask, then: Are there any experimental or observational challenges with which the classical theory of gravity—namely, Einstein's general relativity—is not able to cope? Compared to electrodynamics, the experimental field for testing general relativity has been considerably less fertile. In Chapter 2 we briefly discussed the solar system tests of general relativity, wherein it was seen to perform admirably. Indeed, the theory has done so well in the few tests which have been available, that, on empirical grounds, it is hard to make a case for quantizing it. There are no unexplained gravitational phenomena which *force* the theoretician to come up with a quantum theory of gravity in order to understand them. This situation is unlike that prevailing at the turn of the century with respect to electromagnetism, when observations of spectral lines, the photoelectric effect, and black-body radiation pointed to grave inadequacies in the classical framework.

There are other things, however, which make the quantization of gravity necessary. For example, if we are to succeed in constructing a unified theory of physical interactions, we must quantize gravity in order to bring it into line with other interactions. There is no way one can unify a purely classical theory with theories which are quantized. Even aside from the desire for unification, the theoretician wants to be able to quantize gravity, because of his firm belief that the ultimate description of nature is a quantum one.

Big bang cosmology supplies another motivation for quantizing gravity. If we examine any finite volume of the universe and trace its history back in time, we discover that sufficiently close to $t = 0$ (the big bang epoch), it becomes so small that the classical description breaks down. There is a general rule which tells physicists whether classical laws are adequate to describe the physical systems they are studying. The rule is simply stated: the 'action' describing the system must be large compared to $\hbar$, if the classical description is to be trusted.

The 'action' in the above sentence is the one which classical physicists use in deriving their laws of physics from the

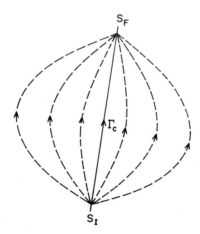

Fig. 6.5. The path-integral approach to quantum mechanics allows the system to proceed along any of the geometrically permissible paths from the specified initial state S_I to the specified final state S_F. In the limit of classical mechanics, a unique path, Γ_C connects S_I to S_F.

so-called principle of least action. The adjective 'least' is in fact a misnomer; the correct word is 'stationary'. The requirement that action be stationary and how it helps in arriving at classical laws of physics may be understood as follows, with the help of Fig. 6.5.

Imagine a physical system which evolves from a given initial state, S_I, to a final state, S_F. How it does so is determined by a law of nature yet to be discovered. If we do not know the form of this law beforehand, we may think of the evolution proceeding along any one of a large number of pathways, which we may denote by a set $\{\Gamma\}$. Our situation at this stage is like that of a tourist in a new city who wishes to go from his present location to some other one. On his map he sees a maze of streets, and does not know the best route to his destination. A local guide, familiar with the traffic restrictions, flow patterns, and distances could help him choose the route which would involve the least bother. The action function, $\mathscr{A}$, is a physical quantity which serves as a guide to the theoretical physicist. The form of $\mathscr{A}$ is usually guessed, with the help of symmetry principles and, of course, intuition.

Once the form of $\mathscr{A}$ is known, it can be computed along any of the available paths. The values of $\mathscr{A}$ so computed will naturally differ from path to path. There may be a particular path, Γ_C, however, with the property that if we make a small change from $\Gamma_{C'}$, the computed action does not change. This is what is implied by the adjective 'stationary', and the path so chosen is the one followed by nature.

The principle of stationary action was first invoked by W. Hamilton in 1824–32 to derive the laws of motion of a complex dynamical system. Although the laws of motion *per se* had already been discovered by Isaac Newton in the seventeenth century, Hamilton's method provided an elegant way of arriving at them, and it proved to be applicable to other branches of physics, where the laws were not already known. It also helped in providing a general principle which seems to govern such diverse laws as those of mechanics, Maxwell's electromagnetic theory, and Einstein's general relativity.

The action, $\mathscr{A}$, also provides a convenient bridge between classical and quantum theory. Note first that the rule of stationary action picks out a unique path, Γ_C, which agrees with the deterministic picture of classical physics. Suppose we now introduce a fuzziness into this picture, by adding a new rule to the effect that we are not in practice able to distinguish between paths whose action differs by quantities of the order $\hbar$. Clearly this limitation will not affect classical physics as long as $\mathscr{A}$ is large compared to $\hbar$. But if $\mathscr{A}$ turns out to be comparable to $\hbar$ along most (or all) paths, then the stationary property loses its significance. The unique track of evolution from S_I to S_F, Γ_C, now gives way to a large number of alternative tracks of evolution, and the best we can do is to assign probabilities to the various evolutionary routes. This is the way quantum physics operates. The same action can therefore lead to different rules of behaviour in the quantum domain (where $\mathscr{A}$ is comparable to $\hbar$) and the classical domain (where $\mathscr{A}$ is large compared to $\hbar$).

By means of our tourist analogy, we may contrast these two

domains in this way. The classical domain corresponds to the situation in which fast motorways more or less unequivocally specify the best route between two major cities. The situation inside a city may be more confusing, with its crowds, one-way streets, parking restrictions, and so on. Here there may not be a unique best way to go from one destination to another: all available routes may be more or less equally good (or bad!). This corresponds to the quantum domain.

Let us now go back to the big bang universe, whose dynamical behaviour we have so far determined by classical rules—namely, by the general theory of relativity. As we just mentioned, this theory is derivable from a stationary action principle; it was D. Hilbert who first demonstrated this elegant result in 1915, soon after Einstein wrote down the equations of general relativity heuristically. If we use the Hilbert action formula and apply it to the classical Friedmann model, we discover that classical laws cannot be considered reliable for t < t_P. In other words, if we continue our adventurous probing of the past of the universe, we come to a natural barrier at the Planck epoch, beyond which our classical picture breaks down.

This is where we find the strongest motivation for studying quantum gravity. Without quantizing gravity, we are not able to say what the universe was like before the Planck epoch. The very basic question of the origin of the universe belongs to this pre-Planck era. We end this chapter by highlighting the difficulties of working out the physics of this era, and the modest successes achieved in spite of them.

Conceptual problems of quantum gravity

Let us enumerate the major difficulties of visualizing the working of quantum gravity at this universal epoch.

First, what is the role of the observer in this situation? Most of the development of quantum theory, in particular its foundational aspect, is linked with the notion of measurement of a microscopic system by an observer who inevitably disturbs the system while measuring it. The quantum uncertainty principle, the jumping of the system from one state to another, the probability of finding it in one of the many possible states, all developed out of this observer–system relationship. How are these notions adapted to the situation wherein the entire universe, including the observer, is a microscopic system?

A second difficulty arises from the general relativity viewpoint. As we saw in Chapter 2, general relativity is different from other physical theories in that it interprets the force of gravitation as a geometrical effect. Indeed, so effectively is gravity removed as a force by the notion of curved space–time that Newton's first law of motion is supposed to hold even if gravity is present. Uniform motion in curved space–time thus replaces the usual concept of accelerated motion under force. So if gravity as a force is eliminated, what exactly are we supposed to quantize?

Clearly, the answer to this question is not trivial. For if we argue that the quantity to be quantized is space–time geometry, then we are talking about an entity which is altogether different from the entities involved in other quantum field theories. For example, when Maxwell's field theory is quantized, the space–time in which the quantization is carried out remains unaltered. Indeed, all the successes of QED to date are in flat Minkowski space–time. In quantum gravity the very process of quantization affects the geometrical structure of space–time, and this generates conceptual problems at all levels.

At the very elementary level the difficulty is highlighted in Fig. 6.6. Consider two events A and B so located in space–time that A causally affects B. This implies that B lies within the future light-cone of A. However, light-cones are determined by space–time geometry, and a change in the latter would in general change the former. Thus it could happen that in the new geometry obtained by a quantum transition, B is causally disconnected from A, a result which has no analogue in ordinary quantum field theory. At a more technical level this breakdown of causality can affect the quantum parameters which describe how quantum effects of various physical fields propagate across space–time.

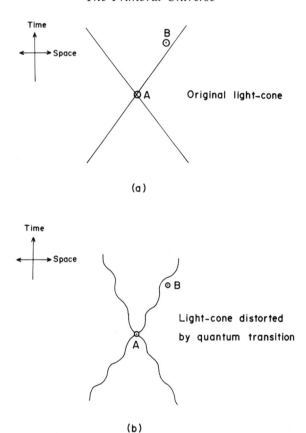

Fig. 6.6. The distortion of light-cones and causal connections in quantum transitions of space–time geometry. In (a) the event A causally affects B, whereas in (b) the two are causally disconnected because of a new geometry.

Technical problems of quantum gravity

In spite of these serious conceptual problems, one may proceed to quantize Einstein's general relativity in a purely formal way, with the hope that as concrete results emerge from the formalism, the conceptual problems may become easier to understand. However, even here, purely technical problems hinder progress towards concrete results. I mention only two such problems.

The first springs from the non-linearity of general relativity. The field quantities, even if they are taken as the space–time metric or the quantities specifying space–time curvature, do not obey the superposition principle so essential to the usual formulation of field theory. A very simple example will illustrate this difficulty. In electromagnetic theory the superposition principle holds. The total field present when there are two electric charges is simply the sum of the fields when each of the charges is present alone. In general relativity we know that an isolated spherical mass generates the Schwarzschild space–time (see Chapter 2). What is the space–time geometry when two masses are present? An exact solution of this problem is not yet available; but it is certainly not the case that we can simply superpose two Schwarzschild space–times.

Thus new complicated techniques are needed to formally quantize gravity *exactly*. What if we compromise on exactness and linearize the theory to the situation where gravitational effects are small? Classically, we can then use the superposition principle; but our 'weak field' approximation cannot be applied to quantum cosmology, where gravitational effects in the pre-Planck epoch are enormous. Nevertheless, even the weak field theory has its own difficulty, in that it is non-renormalizable. In Chapter 4 we saw why renormalizability is considered an essential requirement in quantum field theory.

In view of such difficulties it is not surprising that, despite considerable intellectual gymnastics, formal approaches to quantum gravity—there are more than one!—have failed to deliver the goods in the form of concrete applicable results. There are some physicists who believe that the task will never succeed, and that the answer may be via supergravity theories. But it is early days yet to assess those theories.

Conformal Quantization

To end this chapter on a positive note, I will outline briefly an approach to quantum gravity which is less ambitious than the usual 'let us deliver a complete theory of quantum gravity' type of approach. It has the merit, however, of dealing with questions of direct relevance to quantum cosmology.

Recall the conceptual problem of the formal approach: namely, that in the very process of quantization the light-cone structure of space–time changes. Is there any exception to this rule? Are there any non-trivial changes of space–time geometry which *preserve* the light-cones as they were? The answer is yes. Such changes are called 'conformal transformations'. We introduced them in Chapter 4 in describing Weyl's gauge theory.

To obtain one geometry from another by a conformal transformation, we simply introduce a conformal function, Ω, which scales uniformly all the lengths measured at any point in space–time, while preserving all the angles. The lengths and angles refer of course to measurements of time as well as space. Such a transformation would be trivial if Ω were to have the same value at all points of space–time, since the so-called change could simply be ascribed to a uniform change in the unit of distance measurement throughout the universe. For a non-trivial change, Ω is expected to vary from point to point. The distinction between the two cases is the same as that between the global and local transformations described in Chapter 4.

The important point is that even under *local* changes of Ω, the light-cone structure is preserved *globally*. Thus, provided we stick to quantizing only the conformal degree of freedom of the space–time geometry, we do not run into any problems of causality breakdown. The other advantage of such a restriction is purely computational: it turns out that the full quantum theory of Ω can be worked out in spite of its non-linearity. Thus no approximation is involved. We shall refer to this restricted quantization as 'conformal quantization'.

It is immediately obvious that of the various degrees of freedom which might be used to describe the range of space–time geometries, the conformal degree is the one most relevant to our cosmological problem. The expansion of the universe is characterized at the simplest level by the change of scale of its spatial volume. Such a change can be described by Ω. In fact, in order to describe the range of Friedmann–Robertson–Walker space–times, we need only the conformal degrees of freedom. This can also be seen from the exact mathematical result that all such space–times are obtained by conformal transformations of Minkowski space–time. That is, by introducing a suitably varying scale factor Ω into the Minkowski space–time, we can reproduce the geometry of the Robertson–Walker type. The singularity common to Friedmann models is then seen as the vanishing of Ω at a particular epoch. We shall use this result explicitly later.

So if we wish to capture the essence of quantum cosmology *vis-à-vis* the behaviour of the very early universe soon after the singular epoch, quantizing Ω is our best bet. Can we expect conformal quantization to lead us to radically different results?

Our expectation is guided by a more familiar and considerably simpler example from quantum theory of the hydrogen atom. The classical picture of the hydrogen atom, that of an electron circling round a proton, was seen to be untenable. For the electron in this picture is for ever changing its direction as it goes round; it is therefore being accelerated, and, by Maxwell's classical theory, should radiate energy. The energy loss would result in the electron spiralling inwards and falling onto the proton in a time of the order of 10^{-23} s! Obviously the stability of the hydrogen atom gives the lie to such a scenario. To capture the essence of the quantum picture, we proceed first to quantize the most relevant degree of freedom in the motion of the electron—namely, its distance from the nucleus, r. By quantizing r alone, we discover that the electron can exist stably in a 'stationary' orbit with a characteristic size $r \cong \hbar^2/me^2$, where m and e are the mass and charge of the electron respectively. This difference between the classical and quantum pictures is illustrated in Fig. 6.7.

Notice that even though we have not quantized the angular motion of the electron, we have been able to derive the important result that, quantum-mechanically, there is no catastrophic end to the electron's motion. A complete quantized theory of the hydrogen atom gives us many more

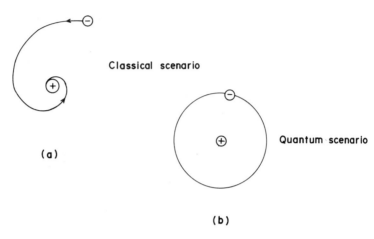

(a)

Classical scenario

Quantum scenario

(b)

Fig. 6.7. The classical electron winds its way into the nucleus of the hydrogen atom (a). The quantum electron stays in a stationary state 'well away' from the nucleus (b).

stationary states for the electron, of course, of which the one above obtained by quantizing r alone turns out to be the stablest. Based on this partial quantization example, our expectation is that conformal quantization should tell us the answer to the infinite-dollar question as to whether the space–time singularity is avoided in quantum cosmology.

This was the hope which led me to explore conformal quantization further, in spite of a possible criticism that it gives only a partial picture of what may happen in quantum cosmology. Summarized below are some interesting conclusions of this work, part of which was in collaboration with Thanu Padmanabhan.

First, notice that by introducing conformal transformations of the Friedmann model, we generate new space–times whose geometries do *not* satisfy the classical Einstein equations. This is to be expected, since quantum physics inevitably opens out a vista far wider than that of classical physics. The question is: Of the new models of the universe so generated, are there some without any singularity? The answer to this question is yes. There are non-singular, as well as singular, models of the

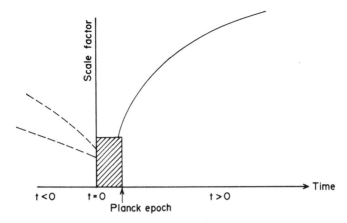

Fig. 6.8. A non-singular 'beginning' to a Friedmann model. In the shaded, quantum era the scale factor was probably not zero, so the universe had a history before $t = 0$ (shown by dashed extensions in $t < 0$ region).

universe in quantum conformal cosmology. However, because we are in the quantum regime, we can no longer pinpoint a definitive history of the universe: we can talk only in terms of probability. Thus, instead of the definitive statement of classical cosmology that the universe had a singular origin, we can at best ask about the probability that the universe had a singular origin. The clear answer which emerges from conformal quantization is that this probability is vanishingly small. In other words, it is extremely unlikely that the universe originated in a big bang. Fig. 6.8 describes this new situation brought about by quantum theory. This conclusion has been generalized to conformal transformations of other (non-Friedmann-like) solutions of relativistic cosmology. Here, too, non-singular models predominate over singular ones.

This answer is quantitatively similar to that regarding states of the hydrogen atom, and it prompts one to ask whether 'stationary states' could exist for the universe. Padmanabhan has been able to demonstrate that such states indeed exist within the framework of conformal quantization, and that the

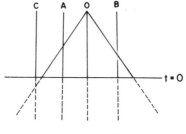

Extension to t<0

Fig. 6.9. The light-cone of O in the absence of a singularity at $t = 0$ can now be extended back to the region $t < 0$, since the universe existed then. So there is no particle horizon at O. Particles like C are also able to influence O causally.

characteristic scale associated with stationary models is, not surprisingly, the Planck length:

$$L_P = \sqrt{\frac{G\hbar}{c^3}} \cong 1.6 \times 10^{-33} \text{ cm.}$$

If we regard these results as having eliminated the singularity problem of classical cosmology, we can proceed further and argue that in all probability the horizon problem described in Chapter 5 has also been removed. For, as shown in Fig. 6.8, we now appear to have a universe with a history extending well past the $t = 0$ epoch, and there is no abrupt termination of the past light-cones which gave rise to particle horizons. Fig. 6.9 shows that light-cones can now be extended to epochs earlier than the classical epoch of the big bang.

It is interesting to find that conformal quantization also resolves the flatness problem. Suppose we argue (as many theoreticians do) that the universe itself arose as a quantum fluctuation of a pure vacuum, from an initial state described by the empty Minkowski space–time. Is there any particular mode into which such a transition would take the universe? Padmanabhan used conformal quantization to find the answer. Recall that the Robertson–Walker space–times have geometries which are conformal transformations of Minkowski geometry. So these will be among the candidates for the final state of the universe. Again one can calculate probabilities for the various transitions. The outcome is that by an overwhelming probability the final state turns out to be the $k = 0$ Robertson–Walker space–time! Thus one can assert that if the universe evolved out of the empty Minkowski space–time by quantum conformal fluctuations, then almost certainly it would go into a flat ($k = 0$) Robertson–Walker model.

It is clear, therefore, that the problems of horizon and flatness which gave rise to the inflationary scenario are solved in the much earlier quantum era. In addition, the problem of space–time singularity, for which inflation does not offer a solution, also gets solved by conformal quantization.

What has been achieved here *vis-à-vis* a full theory of quantum gravity may be compared with what the simple-minded approaches to quantum mechanics which prevailed in the first two decades of this century did *vis-à-vis* the more complete descriptions of Schrödinger, Heisenberg, and Dirac in the 1920s. Whatever the ultimate assessment of conformal quantization, it at least demonstrates that the pre-Planck era may be a very important phase in the history of the universe.

7

The Universe in Retrospect

I think it is very unlikely that a creature evolving on this planet, the human being, is likely to possess a brain that is fully capable of understanding physics in its totality. I think this is inherently improbable in the first place, but, even if it should be so, it is surely wildly improbable that this situation should just have been reached (now) . . .

Fred Hoyle

'Follow the crowd and you will find the right way', says an old Sanskrit verse. The advice works for someone heading for a sports event or a tourist attraction. It is based on the assumption that the crowd knows where it is going. But it is an assumption which has not always worked in science, especially in frontier areas. The epicyclic theory of the Greeks and the concepts of phlogiston, perpetual motion, an all-pervading aether, and so on are examples where crowds, including great scientific brains of the time, were led astray. Even in comparatively recent times in the long history of astronomy, the majority view proved to be wrong on such important issues as the sun's location in the Galaxy and the extragalactic nature of diffuse nebulae like Andromeda.

It is therefore with some caution that we view the recent flurry of activity on the particle physics–cosmology frontier. The interaction between physics and astronomy has proved mutually beneficial. Astronomical phenomena, however weird they look at first sight, gain in comprehension when described as consequences of known physical laws. Likewise, the physicists' confidence in the validity of these laws is bolstered considerably by seeing their applicability to cosmic settings which far exceed the range of possibilities in terrestrial laboratories. But for the interaction to be successful, at least one of the two disciplines must rest on secure grounds.

For example, the early universe calculations of Chapter 3 are considered well founded, because the physics of particles on which they are based has been well tested, even though astronomical observations of the universe at the age of 1 second to 3 minutes are non-existent. Similarly, although controlled nuclear fusion has not yet been achieved in the laboratory, detailed observations of stars generate confidence in the physical theory of fusion of nuclei.

When it comes to the work described in Chapters 4–6, however, we are dealing with speculations on *both* fronts. There are no direct astronomical observations of the very early universe; and the physical theories used to describe it have not been tested independently for validity. It is because we are matching one speculation against another that we need to be cautious about the physical reality of the entire scenario. To take a medical analogy, we are using untested medicines for diseases which have not only been diagnosed poorly, but whose very symptoms are rather tentative.

As we pointed out in Chapter 5, the sole justification for this procedure lies in the attempt to explain the present state of the universe as a relic of the very early epochs. The tail-eating snake shown in Fig. 7.1 illustrates the present scenario. But efforts in this direction have been only modestly successful, and certainly do not justify the confidence sometimes publicly expressed by leaders in the field, that with GUTs and SUSY round the corner, the end of physics is at hand, as is the solution of the ultimate cosmological problem.

The fact that high energy particle theories currently in fashion generate unwanted relics in the very early universe, relics which have to be got rid of by additional scenarios, may remind the cynic of the old story of the artist who displayed an empty canvas on a wall, claiming that it depicted a cow eating grass. Why no grass? Because the cow had eaten it. Why no cow? Because the cow had no interest in remaining there once the grass was gone. The *absence* of monopoles, domain walls, seeds of galaxies, and the rest is confidently taken to support a

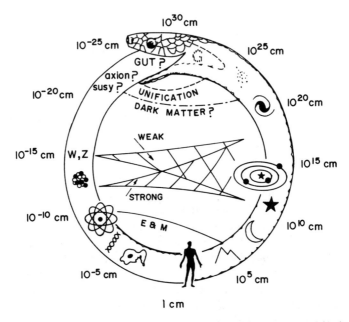

Fig. 7.1. The famous Glashau's snake, illustrating the way physical interactions play a crucial role in determining the state of the universe at different epochs. This picture of standard cosmology may be compared with Fig. 1.1.

number of hypotheses regarding the unification theory and the very early universe. Nowhere else in physics have null observations been used to claim so much that cannot be otherwise tested.

The cynic may also remind us that the short time span of $\sim 10^{-32}$ s provided the only opportunity for the gamut of GUTs, SUSY, K–K theories, and quantum gravity to play their crucial role. We therefore do not have the criterion so precious to science—namely, the repeatability of experiments. Even in astronomy this situation is unique; for in other applications of physics to astronomy, to phenomena such as stellar evolution, stellar nucleosynthesis, galactic structure, quasars, radio sources, and so on, there is a continuing opportunity of mutual feedback between theory and observations.

For these reasons, while the route chosen by the crowds may look very scenic and may well lead to the desired destination (namely, the end of physics and cosmology by the end of the twentieth century), it may be worthwhile to take note of a few isolated byways which have attracted limited attention. We therefore end this cosmic odyssey by taking short detours along three such byways in cosmology.

The Steady State Theory

Proposed in 1948 by Hermann Bondi, Tommy Gold, and Fred Hoyle, the steady state theory is in many ways the opposite of big bang cosmology. Instead of an evolving universe with a finite age and an explosive origin, we have a steadily expanding universe without a beginning or an end, in which creation (of matter and energy) is going on continually.

The motivation which led Bondi and Gold to the steady state concept sprang from the speculative nature of physics and cosmology in the early post-big bang era. Theoretical physicists in the 1940s were not willing to speculate about the very early universe. The reason, as Bondi and Gold perceived, was that the physical conditions and the laws governing them in those early epochs would be markedly different from those being studied 'here and now'. Rather than get caught up in untestable speculations, they proposed an alternative scenario which could readily be tested, a scenario which rested on the premiss that the overall physical conditions in the universe and the laws governing them do not change at all.

In a sense the Bondi–Gold notion of an unchanging universe is a logical extension of the cosmological principle of Chapter 2. The cosmological principle ensures that the state of the universe and the laws of physics influencing it are identical here and elsewhere *at the same cosmic time*. Astronomical observations, on the other hand, tell us about far-away regions in the universe not at the *present* cosmic time, but at *earlier epochs* when the light arriving now left them. So to guarantee a meaningful comparison of observations with theory, we require the universe and its physical laws to be unchanging with time as well as space. This is the essence of the 'perfect cosmological principle' (PCP) which led Bondi and Gold to the steady state theory.

An unchanging universe does not necessarily mean a static universe. In fact, in a static, infinitely old universe, physical systems would reach a thermodynamic equilibrium, which is not the case with the present universe. The alternatives provided by the PCP are either a steadily contracting universe or a steadily expanding universe, of which the former is again ruled out by thermodynamics. So we are left with the alternative that the universe has been expanding steadily. Notice that this deduction was arrived at without any dynamical theory like general relativity.

The deductive power of the PCP also tells us that the space–time geometry of the steady state model is described by Robertson–Walker space–time, with $k = 0$ and the scale factor, $S \propto \exp H_0 t$, with H_0, the Hubble constant, truly a constant with respect to the cosmic time, t. It is easy to see, for example, how the above expansion rate comes out of the PCP. The principle requires that in an unchanging universe the Hubble constant, S/S, must not change. Only an exponential function satisfies this criterion. Later we will return to the fact that the steady state space–time is exactly the same as the de Sitter space–time of Chapter 2, the space–time later used by the inflationary models.

It was emphasized by Bondi and Gold that the PCP combined with local observations of the universe tells us everything observable about the universe. Their main stress was on the testability of the PCP, which made unequivocal claims about the large-scale structure of the universe, and was therefore much more vulnerable to observational constraints than the evolving Friedmann models, which ascribed the bulk of the present observable state of the universe to speculative, very early epochs.

In the end, the strongest observational challenge to the PCP and the steady state universe came from the discovery of the microwave background. This background must be unchanging, according to the PCP, and it must be continually regenerated as the universe expands. Its origin must therefore be entirely astrophysical in nature. But how can this be? A related problem arose *vis-à-vis* the abundance of light nuclei, especially helium and deuterium. Since there was no hot epoch in the past in the steady state universe (because there is none now), these elements must also be produced continually now. But how?

These questions swung the balance of opinion heavily towards the big bang model in the 1960s, and it has remained there ever since. To sustain the steady state idea, one must show how light nuclei and the microwave background can be regenerated without recourse to hot universal conditions. Although considerable progress has been made in the 1970s and the early 1980s towards this goal, a detailed solution is still to be obtained. The clue to the solution is provided by the circumstance that if *all* the observed helium were somehow synthesized in stars, the amount of starlight so generated, when converted to black-body radiation, would give the microwave background of about 3 K temperature. The success of the solution depends on demonstrating helium production on this scale in stars and the process of thermalization of the resulting starlight. The steps in such a scenario are shown in Fig. 7.2.

Hoyle's approach to the steady state model was not via a deductive principle like the PCP, which has its own limitations in not giving a manifest connection between the state of the universe and basic physical laws. Hoyle was motivated by wishing to provide a physical theory for the creation of matter. In the standard Friedmann scenario the creation of the universe at the big bang is a singular event which cannot be studied as a physical phenomenon. What if the matter observed in the universe were not created in 'one go' at one epoch, but is being continually created at all epochs? The standard field-theoretic picture of particle creation does not permit this scenario in relativistic cosmology. Hoyle therefore proposed a modification of the right-hand side of Einstein's

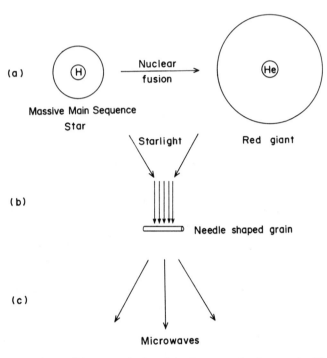

Fig. 7.2. A possible scenario for the microwave background without a hot big bang. Massive stars burn out quickly and produce all the helium observed in the universe (a). Most of this starlight is absorbed by intergalactic dust in the form of long, needle-shaped graphite grains (b), and is re-emitted in microwaves (c). It can be shown that the resulting background will have the observed temperature.

field equations by adding an extra scalar field called the 'Creation field', or simply the 'C-field'. Later, in 1960, Maurice Pryce provided an elegant mathematical formulation for the C-field, using an action principle.

The C-field has the unusual feature that it has *negative* stresses and *negative* energy. Negative energy sources, when coupled with Einstein's gravity, generate repulsion, which is why the universe continually expands in an accelerating fashion. (By contrast, expansion in the Friedmann models was generated by the initial big bang, and is steadily slowing down.) The negative energy and stresses also allow for continuous creation of matter *without* violation of the sacrosanct law of conservation of matter and energy. This is where particle creation via the C-field differs from that via a standard positive-energy field like the electromagnetic field.

The steady state model emerges as the simplest solution of these modified equations. Although the uniqueness of solution which follows from the PCP is absent in Hoyle's approach, there is a direct relationship between the parameters of the solution and the parameters of the physical theory. Thus Hubble's constant, H_0, can be related to the gravitational constant and to the coupling constant of the C-field to gravity. The mean density of matter in the universe is given by

$$\rho_0 = \frac{3H_0^2}{4\pi G}.$$

And matter is being continually created at the rate of $3H\rho_0$ per unit volume. For $H_0 = 100$ km s^{-1} Mpc^{-1}, this rate works out at about 4×10^{-46} g cm^{-3} s^{-1}.

Although the steady state model as given by the PCP addresses the two problems described earlier, the above field-theoretic approach and its wider range of cosmological models contain many features which are popular today among supporters of the inflationary scenario. For this reason it is of interest to describe briefly how the steady state theory anticipated some features of the present work on the very early universe.

Before going on to C-field cosmology, I should highlight W. H. McCrea's ideas of 1951 on why and how the steady state universe expands. Using the de Sitter space–time of the steady state model, McCrea found that the framework of general relativity demands *negative* stresses in the energy–momentum tensor describing the physical contents of the universe. Thus, according to McCrea, the universe expands, and matter is continually created, because of the negative stresses which, he conjectured, arise from what goes on in the vacuum. In the early 1950s particle physics had not acquired the sophistication necessary to explain how negative stresses could be generated by a vacuum, and McCrea's ideas were not taken very seriously. Three decades later, we appreciate how prescient they were.

What form does the newly created matter take?

Back in the mid-1950s Geoffrey Burbidge and Fred Hoyle considered the possibility of an equal production of matter and antimatter so as to preserve the baryon number. But this idea ran into difficulties, because the annihilation of matter and antimatter produced gamma radiation in profusion, leading to an unacceptably high background of such radiation. In 1958 Gold and Hoyle suggested that the matter created was in the form of neutrons which subsequently undergo beta decay: $n \rightarrow p + e^- + \bar{\nu}$. It was Gold's and Hoyle's view that the energy released in this process would generate a large kinetic temperature in the electron population, and that this in turn would eventually lead to large pressure gradients across distances of 50–100 Mpc. These gradients would trigger the formation of large groups of galaxies which (as was known even then!) could not be formed purely gravitationally.

This remarkably prescient model anticipated the concept of baryon non-conservation and structure at the level of super-clusters of galaxies. The former was unacceptable to the particle theorists of the 1950s, however, and the latter to astronomers, who had barely begun to notice inhomogeneities on scales of 50–100 Mpc. In the event, the 'hot universe' idea collapsed, because 'hot' electrons were found to generate too high a background of X-rays.

These examples illustrate the advantages of the steady state idea *vis-à-vis* current very early universe scenarios. The latter employ extrapolations of the laws of physics which can never be tested independently of cosmology, and which involve speculations about physical processes that took place in the remote past, when the universe was radically different from what it is now. In the steady state theory, on the other hand, we talk about physical processes which are supposed to be going on in the universe *now*, rather than in the very remote past. Thus its predictions, like those just mentioned, can be tested directly by observations, a process which physicists have traditionally followed in astronomy.

C-field cosmology provided a consistent field-theoretic picture of the creation of matter and the steady state expansion. Insertion of a field of negative stresses and negative energy into the general relativistic framework removes the singularity problem. For the various singularity theorems proved during the 1960s, which led to the conclusion that a big bang type of singular epoch was inevitable according to general relativity, are based on the assumption of 'positivity of energy'. The C-field interactions of negative stress and negative energy provide a loophole in the arguments of these theorems.

When Hoyle and I started exploring the physical properties of the C-field in the early 1960s, we noticed another elegant property. Suppose that in the remote past the universe was not in the highly regular steady state that it is in now. No matter what state it was in, the C-field would have caused matter to be created in such a way that the universe would have begun to attain regularity. Imagine by way of analogy an electric circuit containing transient currents, into which a battery is introduced. The steady current coming from the battery will eventually dominate, with the transient currents dying out. Likewise, the regularity brought in by the newly created matter begins to assert itself on time-scales of the order of H_0^{-1}. Thus we understand why the universe reaches such a highly symmetric state.

In retrospect, this is no different from the concept of a chaotic beginning leading to regularity which we described in Chapter 5. This is hardly surprising, since the dynamical picture of how inhomogeneities are damped in the inflationary phase is identical with that in C-field cosmology.

The very notion of exponential inflation and the subsequent generation of a Friedmann bubble which form the central features of the inflationary scenario seem to have been anti-cipated by C-field cosmology back in 1964–6, although the 'physics' of the two models is different. While studying the

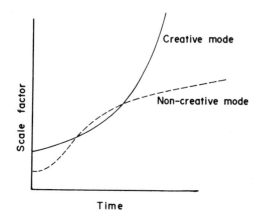

Fig. 7.3. The expansion rates of the universe for the two types of C-field solutions.

C-field-generated solutions of Einstein's equations, Hoyle and I found that basically there were two types of solutions. In 'source-free' solutions the C-field did not generate any matter at all. Its effect on the dynamics of the universe died out rapidly as the universe expanded. Its presence made only a marginal difference, therefore, except in one situation which I will outline shortly. Thus, by and large, the source-free solutions were similar to Friedmann-type solutions for the three values of the curvature parameter $k = +1, 0, -1$. They differed from the Friedmann models only when the universe was very dense. In this case the C-field effects grew very rapidly, generating huge negative stresses which did not allow the universe to be singular. In short, the source-free models turned out to be non-singular versions of the Friedmann models.

The solutions 'with sources', on the other hand, were somewhat unusual and implied creation of matter. These are the solutions which eventually led to the highly regular steady state universe. Both these solutions are illustrated in Fig. 7.3.

Now imagine that in a universe given by the second type of solution, the process of matter creation spontaneously stops in a small volume of space. Theory then tells us that this volume will subsequently expand as in the 'source-free' case—that is, like a Friedmann universe. Thus we have a Friedmann bubble growing up inside a denser steady state universe. Suppose that our observable universe is confined to such a bubble. We will then find from local observations that we live in an 'unsteady' universe. Only by probing the remote past could we hope to discover the steady universe from which the bubble evolved.

What can we say about the universe which lies beyond the bubble? Hoyle and I visualized the possibility that the coupling constant of the C-field was so strong that the density of the outer steady state universe would be several orders of magnitudes higher than the so-called closure density of the Friedmann bubble at the present epoch.

The reader cannot fail to notice the similarity between this picture and the inflationary model which came fifteen years later. Because the bubble in C-field cosmology grows out of a steady state universe, it corresponds to the $k = 0$ Friedmann model. Thus the flatness problem is resolved. The singularity and horizon problems are already resolved, since the external steady state universe does not have a singular beginning or particle horizons.

In retrospect, one can only argue that steady state cosmology came on the scene much too soon, before physicists and astronomers were ready for it. The treatment meted out to the theory reflects a fact of life in the sociology of science: that theoreticians tend to ignore observational realities until they have a theory for them, and that observers suspect theoretical ideas unless they already have observations to match them.

The Anthropic Principle

The Copernican revolution was the first step in man's dethronement from the 'centre' of the universe. The steady erosion of man's privileged status reached its ultimate stage in the cosmological principle (or the PCP if the steady state model had turned out to be right). In the homogeneous isotropic universe all fundamental observers have the same status. Man in his galaxy is just one of them. Since the trend towards democratization of the universe started with Copernicus, we may call this ultimate concept of cosmic equality the 'Copernican principle'.

A reaction to the Copernican principle, however, was initiated by Robert Dicke in 1961, with the so-called anthropic principle. In general terms this principle amounts to the statement that the universe is the way it is because we are here to observe it. By 'we' is implied the typical human observer who has attained a certain level of intelligence in the course of the evolution of life in the universe. Had the universe been different in its structure and evolution, it would not have been possible for such human observers to evolve to their present stage. This is what the anthropic principle is all about. It is a deductive principle which, one hopes, might narrow the ranges of the parameters in the physical theories and the initial conditions, and lead to a unique model of the universe.[1]

Consider the following application of this principle proposed by Brandon Carter to show how the magnitude of the gravitational constant, G, turns out to be strongly related to our existence. To follow Carter's argument, let us consider the stars which derive their luminosity from nuclear fusion of hydrogen into helium. This process goes on steadily for a considerable period, which constitutes the bulk of the life span of a star. Now in such a state there is a definite relation between the luminosity, L, and surface temperature, T, of a star. In a logarithmic plot of L against $1/T$, such stars lie on a narrow band called the 'main sequence', which extends from high values of L and T, at the so-called blue end, to low values of L and T, at the red end of the sequence, the names corresponding to the dominant colour in the star's radiation. Fig. 7.4 describes how a typical group of stars appears on the (L, T) diagram.

Now, the crucial parameter which determines where a star is on the main sequence is its gravitational mass. Stars of large mass are at the blue end and are called 'blue giants', whereas those of low mass, 'red dwarfs', are at the red end. It is also the case that the hydrogen fusion reaction goes fast in high-

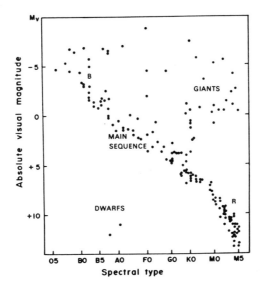

Fig. 7.4. In this Hertzsprung–Russell diagram for the nearest stars, the blue giants are at end B, and the red dwarfs at end R of the main sequence. The absolute visual magnitude is a logarithmic measure of the luminosity, and the spectral type is a logarithmic measure of the reciprocal of the surface temperature of a star.

mass stars and slow in low-mass stars. Red dwarfs therefore remain on the main sequence considerably longer than blue giants.

Now consider the origin of life and its evolution to an intellectually advanced state. Suppose that this event occurs on a planet going round a star, and that the energy for sustaining life comes from the star. This evolutionary process requires a certain amount of time, as well as sufficient radiation. Blue giants have the radiation but not the time, whereas red dwarfs have the time but not the radiation. The conclusion? That life originates and evolves only around stars which are somewhere in between these two extremities. The sun, lying in this middle region, satisfies these two conditions.

If we now imagine a universe with a considerably larger gravitational constant than the one we actually have, the gravitational masses of all stars in it would be effectively higher. Thus the stars in that universe would tend to be like the blue giants in our universe. Similarly, in a universe with a weaker gravitational constant, the stars would be like our red dwarfs. In neither case would life as we know it be possible. Therefore G must take values in a moderately narrow range around its observed value.

The argument as presented above illustrates how the anthropic principle operates. As given here, it is not sharp enough for believers in the principle who would like to demonstrate that values of physical parameters are finely tuned to human existence. Nor is it convincing enough to persuade sceptics who might attack its speculative nature. After all, we still know too little about the formation of planets, about the origin and adaptability of life, about the evolution of intelligence, and so on to be able to conclude definitively 'what would have happened, if ...'.

As for sharpening the Carter-type argument, some progress has been made in bringing other physical constants into the discussion of the anthropic principle—for example, the fine structure constant and nuclear binding. Of course, the ultimate success or failure of this line of reasoning must await a deeper understanding of biology.

Biology and Cosmology

The biological considerations peeping out of the anthropic principle may turn out to be the thin end of a wedge. The question 'How probable is the origin of life in the universe?' has cosmological connotations regardless of whether life exists anywhere beyond the earth. For suppose life started spontaneously on the earth because of the emergence of certain organic molecules under a favourable environment. What is the chance of such an event occurring by accident? The following argument by Hoyle suggests that the chance may indeed be extremely small.

Enzymes play key roles in the interactions of biological

systems with their environment. For example, there are enzymes which are responsible for repairing the damage caused by X-rays and ultraviolet light; there are others responsible for absorption of sugars, for breaking linkages in polysaccharides, and so on. The enzyme, which is made up of a chain of amino acids, has to be specifically constructed to serve as a catalyst in a particular chemical reaction. In other words, an enzyme is not an arbitrarily chosen chain, like a code word chosen at random from a jumble to convey a significant message.

Even estimated conservatively, each chosen arrangement for a particular enzyme is one among many. For example, as estimated by F. B. Salisbury in 1969, at least fifteen amino acids must be correctly ordered in a typical enzyme. With twenty alternatives available at each link of the chain, the number of such randomly made chains is 20^{15}, or nearly 10^{20}. The chance of a particular enzyme being made out of such random links is therefore 1 in 10^{20}. Since there are at least 2,000 such enzymes which play key roles in the behaviour of living systems, the chance of arriving at them by a spontaneous accidental process is $10^{-20} \times 10^{-20} \times \ldots$ (2,000 times) $= 10^{-40,000}$. This absurdly small number highlights the improbability of living systems arising purely by accident.

Francis Crick, in his book *Life Itself*, has indulged in a similar game of probability computation. Looking at proteins as vast chains of randomly assembled amino acids (see Fig. 7.5), Crick computes the chance of a 'correct' chain of 200 links emerging this way to be 1 in 20^{200}—that is, about 10^{-260}. Again, the probability of spontaneous emergence of life by chance is far too small, and one wonders if the big bang universe has had time enough at its disposal to allow such a rare event to take place. The analogies which spring to mind are typified by questions like these: How long will a monkey randomly hitting a typewriter keyboard produce a Shakespearean sonnet purely by chance? Or how long will a spontaneous random assembly of component parts, nuts and bolts and so on, take to produce a jumbo jet?

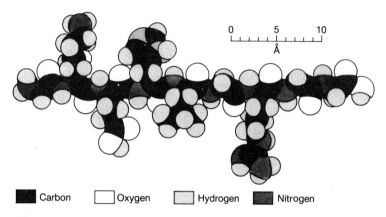

Carbon Oxygen Hydrogen Nitrogen

Fig. 7.5. The polypeptide chain shown here is only nine amino acids long. The unit of length is 1 ångström = 10^{-10} m. Longer chains exist in proteins or enzymes. From Francis Crick, *Life Itself*, p. 86.

It could be argued that the computation of the probability of an event after it has occurred can be misleading. It is also arguable that events that seem entirely unconnected and whose probability of occurrence in a given sequence therefore appears to be too small to be realistic may in fact be connected in a way which we do not yet know. There might seem to be a very small probability that the regular arrangements of electrons in atoms came about by chance. Yet the quantum theory of atomic structure fully accounts for those arrangements as natural ones. Likewise, could these low probabilities in biology be unrealistic because we lack some underlying theory?

However one chooses to look upon these low probabilities of spontaneous generation of life, one is driven to the conclusion that biology contains as yet unravelled information which cosmologists may one day find highly relevant to their search for cosmic roots. Just as questions of the origin of particles and nuclei are seen to be of cosmological importance now, so in future we may find information regarding the origin of living systems to have a bearing on what model we choose to describe the universe. Until such information is forthcoming, our view of the universe, of its present composition and its past history, will necessarily remain incomplete.

Rather than assure the reader that the cosmological problem is all but solved, I prefer to end on this note of incompleteness.

Notes

Chapter 1

1. *Syntaxis*, more commonly known in the translation by the Arabs, *Almagest*. Translated into English by R. C. Taliaferro (Chicago: Encyclopaedia Britannica, 1952).
2. *Aryabhatīya*, ch. 4, v. 9, ed. K. S. Shukla (New Delhi: Indian National Science Academy, 1976).
3. For a detailed description of Kant's ideas, see D. Layzer, *Constructing the Universe* (listed in Further Reading).

Chapter 3

1. One of the early papers included work by Alpher and Gamow. For alphabetical continuity, Gamow included the name of H. Bethe, the nuclear astrophysicist who in 1939 had worked out the model of the sun based on the fusion reaction $4H \rightarrow {}^4He$. The theory reported in the resulting paper became known popularly as the α-β-γ theory.

Chapter 5

1. For details, see G. Steigman, *Annual Reviews of Astronomy and Astrophysics*, 14 (1976): 339.
2. For derivation of the formula, see *Introduction to Cosmology* by the present author (listed in Further Reading).

Chapter 6

1. P. Hut and S. D. M. White, *Nature*, 310 (1984): 637.

Chapter 7

1. For details of the type of arguments presented here, see J. D. Barrow and F. J. Tipler, *The Anthropic Cosmological Principle* (listed in Further Reading).

Further Reading

The following books are either referred to in the text or are listed because they contain more detailed discussions of the topics contained in this book:

J. D. Barrow and J. Silk, *The Left Hand of Creation* (New York: Basic Books, 1983)

J. D. Barrow and F. J. Tipler, *The Anthropic Cosmological Principle* (Oxford: Oxford University Press, 1986)

N. Copernicus, *De Revolutionibus Orbium Celestium*; English translation by Edward Rosen, edited by Jerzy Dobrzycki (Basingstoke: Macmillan, 1972)

F. Crick, *Life Itself: Its Origin and Nature* (London: Macdonald and Co., 1982)

Galileo Galilei, *Dialogo dei due massimi sistemi del mundo* (Florence: Landini, 1632); English translation, *Dialogue on the Two Great World Systems*, by S. Drake (Berkeley and Los Angeles: University of California Press, 1953)

E. Hubble, *The Realm of the Nebulae* (New York: Dover, 1958)

D. Layzer, *Constructing the Universe* (New York: Scientific American Library, 1984)

J. V. Narlikar, *Introduction to Cosmology* (Boston: Jones and Bartlett, 1983)

J. V. Narlikar and T. Padmanabahn, *Gravity, Gauge Theories and Quantum Cosmology* (Dordrecht: D. Reidel Publishing Company, 1986)

I. Newton, *Philosophiae Naturalis Principia Mathematica*, 1st edn. (London: Streater, 1687); English translation, *Mathematical Principles of Natural Philosophy*, by A. Motte, revised by A. Cajori (Berkeley and Los Angeles: University of California Press, 1934; paperback edn., 1962)

S. Weinberg, *The First Three Minutes* (New York: Basic Books, 1977)

Index

Figures in bold indicate locations where a topic is discussed at length.